Inorganic and Organometallic Photochemistry

Mark S. Wrighton, EDITOR

Massachusetts Institute of Technology

A symposium sponsored by the Division of Inorganic Chemistry at the 174th Meeting of the American Chemical Society, Chicago, Illinois, August 31–September 1, 1977.

ADVANCES IN CHEMISTRY SERIES **168**

AMERICAN CHEMICAL SOCIETY

WASHINGTON, D. C.　　1978

Library of Congress CIP Data

Main entry under title:
Inorganic and organometallic photochemistry.

(Advances in chemistry series; 168 ISSN 0065-2393)

Includes bibliographies and index.

1. Photochemistry—Congresses. 2. Chemistry, Inorganic—Congresses. 3. Organometallic compounds—Congresses.
I. Wrighton, Mark, 1949-   . II. American Chemical Society. Division of Inorganic Chemistry. III. Series.

QD1.A355 no. 168   [QD701]   540'.8s   [541'.35]
ISBN 0-8412-0398-9   ASCMC8   168 1–231 78-17616

Copyright © 1978

American Chemical Society

All Rights Reserved. The appearance of the code at the bottom of the first page of each article in this volume indicates the copyright owner's consent that reprographic copies of the article may be made for personal or internal use or for the personal or internal use of specific clients. This consent is given on the condition, however, that the copier pay the stated per copy fee through the Copyright Clearance Center, Inc. for copying beyond that permitted by Sections 107 or 108 of the U.S. Copyright Law. This consent does not extend to copying or transmission by any means—graphic or electronic—for any other purpose, such as for general distribution, for advertising or promotional purposes, for creating new collective works, for resale, or for information storage and retrieval systems.

The citation of trade names and/or names of manufacturers in this publication is not to be construed as an endorsement or as approval by ACS of the commercial products or services referenced herein; nor should the mere reference herein to any drawing, specification, chemical process, or other data be regarded as a license or as a conveyance of any right or permission, to the holder, reader, or any other person or corporation, to manufacture, reproduce, use, or sell any patented invention or copyrighted work that may in any way be related thereto.

PRINTED IN THE UNITED STATES OF AMERICA

Advances in Chemistry Series

Robert F. Gould, *Editor*

Advisory Board

Kenneth B. Bischoff

Donald G. Crosby

Jeremiah P. Freeman

E. Desmond Goddard

Jack Halpern

Robert A. Hofstader

James P. Lodge

John L. Margrave

Nina I. McClelland

John B. Pfeiffer

Joseph V. Rodricks

F. Sherwood Rowland

Alan C. Sartorelli

Raymond B. Seymour

Roy L. Whistler

Aaron Wold

FOREWORD

ADVANCES IN CHEMISTRY SERIES was founded in 1949 by the American Chemical Society as an outlet for symposia and collections of data in special areas of topical interest that could not be accommodated in the Society's journals. It provides a medium for symposia that would otherwise be fragmented, their papers distributed among several journals or not published at all. Papers are reviewed critically according to ACS editorial standards and receive the careful attention and processing characteristic of ACS publications. Volumes in the ADVANCES IN CHEMISTRY SERIES maintain the integrity of the symposia on which they are based; however, verbatim reproductions of previously published papers are not accepted. Papers may include reports of research as well as reviews since symposia may embrace both types of presentation.

CONTENTS

Preface vii

1. Properties and Reactivities of the Luminescent Excited States of Polypyridine Complexes of Ruthenium(II) and Osmium(II) 1
   Norman Sutin and Carol Creutz

2. Light-Induced Electron Transfer Reactions of Hydrophobic Analogs of Ru(bipy)$_3^{2+}$ 28
   Patricia J. Delaive, J. T. Lee, H. Abruña, H. W. Sprintschnik, T. J. Meyer, and David G. Whitten

3. Photochemistry of Metal–Isocyanide Complexes and Its Possible Relevance to Solar Energy Conversion 44
   Harry B. Gray, Kent R. Mann, Nathan S. Lewis, John A. Thich, and Robert M. Richman

4. Photochemical and Photophysical Processes in 2,2′-Bipyridine Complexes of Iridium(III) and Ruthenium(II) 57
   R. J. Watts, J. S. Harrington, and J. Van Houten

5. Roles of Charge Transfer States in the Photochemistry of Ruthenium(II) Ammine Complexes 73
   Peter C. Ford

6. Solution Medium Effects on the Photophysics and Photochemistry of Polypyridyl Complexes of Chromium(III) 91
   Marian S. Henry and Morton Z. Hoffman

7. Photochemical Processes in Cyclopentadienylmetal Carbonyl Complexes 115
   Donna G. Alway and Kenneth W. Barnett

8. Photo-Induced Declusterification of HCCo$_3$(CO)$_9$, CH$_3$CCo$_3$(CO)$_9$, and HFeCo$_3$(CO)$_{12}$ 132
   Gregory L. Geoffroy and Ronald A. Epstein

9. Photochemistry of Bis(dinitrogen)bis[1,2-bis(diphenylphosphino)ethane]molybdenum 147
   T. Adrian George, David C. Busby, and S. D. Allen Iske, Jr.

10. Use of Transition Metal Compounds to Sensitize an Energy Storage Reaction 158
    Charles Kutal

11. Catalysis of Olefin Photoreactions by Transition Metal States 174
    Robert G. Salomon

12. Photocatalyzed Reactions of Alkenes with Silanes Using Trinuclear Metal Carbonyl Catalyst Precursors 189
    Richard G. Austin, Ralph S. Paonessa, Paul J. Giordano, and Mark S. Wrighton

Author Index 215

Subject Index 225

PREFACE

In recent years it has become apparent that significant opportunities now exist for a contribution to the fundamental understanding of the electronic structure and excited-state reactivity of transition metal inorganic and organometallic complexes. This area of research has captured the interest of a number of new contributors and now represents one of the most exciting fields of modern chemical research. The interest and enthusiasm for the field stems from the potential for practical success in photochemical energy conversion, photochemical synthesis, and the initiation of catalytic processes with light. Stimulation comes from the beginning of conscious efforts to convert light to stored chemical energy and to generate catalysts while many of the basic underlying principles are still not completely elaborated. For example, inorganic and organometallic photochemists have not yet demonstrated excited-state analogues of all ground state processes, especially in the area of bimolecular chemical processes. However, in some cases the photochemical approach is the method of choice in laboratory scale syntheses of certain organometallics and inorganic substances.

This volume represents contributions from a number of emerging sectors in the field of inorganic and organometallic photochemistry. Identifying the articles by the key investigator, the first three by N. Sutin, D. G. Whitten, and H. B. Gray deal with recent studies of photoinduced electron transfer reactions in aqueous solution. The papers by P. C. Ford, R. J. Watts, and M. Z. Hoffman represent new advances in the study of primary photoprocesses of water soluble substances. The remaining half-dozen articles concern aspects of the photochemistry and photocatalytic activity of low-valent organometallic systems. Those by C. Kutal and R. G. Salomon deal with Cu(I) photoassisted reactions of olefins with an eye toward energy storage and specific syntheses, respectively. K. W. Barnett and T. A. George discuss primary, but stoichiometric, photoprocesses of metal carbonyl and dinitrogen complexes, respectively. The articles by G. L. Geoffroy and M. S. Wrighton concern the photochemistry and photocatalytic activity of small, metal carbonyl cluster (three metal atoms) complexes.

I am indebted to the contributors of this volume since they have promptly prepared their work. Delay in communicating their timely work is unfortunate, but the lasting qualities of the articles will not be lost in these few months.

Massachusetts Institute of Technology MARK S. WRIGHTON
Cambridge, Massachusetts
March, 1978

Properties and Reactivities of the Luminescent Excited States of Polypyridine Complexes of Ruthenium(II) and Osmium(II)

NORMAN SUTIN and CAROL CREUTZ

Chemistry Department, Brookhaven National Laboratory, Upton, NY 11973

Properties of the luminescent excited states of polypyridineruthenium(II) and -osmium(II) complexes (*ML_3^{2+}) are discussed in terms of the properties of the ML_3^{2+}, ML_3^{3+}, and ML_3^+ ground state species. *$Ru(bpy)_3^{2+}$ and *$Os(bpy)_3^{2+}$ exhibit spectral maxima at ~ 360 nm and 430–460 nm, but the *OsL_3^{2+} lifetimes (~ 9–85 nsec) are much shorter than those for the corresponding *RuL_3^{2+} species (330–1500 nsec). Excited-state potentials for the complexes were estimated from their spectra and ground state potentials as well as from photocurrent measurements at an n-type TiO_2 electrode. The reactivity of *ML_3^{2+} toward outer-sphere electron transfer parallels that of ML_3^{3+} when it undergoes reduction and that of ML_3^+ when it undergoes oxidation. Reactions between *ML_3^{2+} and ground state polypyridinemetal(II) complexes give rise to electron transfer (+1 and +3 ions) or to energy transfer products depending on the pair of complexes used.

The photochemistry of the polypyridine Ru(II) and Os(II) complexes (ML_3^{2+} where L is a 2,2'-bipyridine or 1,10-phenanthroline derivative) has been the subject of much attention in recent years (1–8). Studies of polypyridine complexes are of interest not only in their own right, but also because these systems might find application in solar energy conversion and storage (9, 10, 11, 12, 13). The polypyridine complexes are especially attractive for solar energy applications because of their spectral properties, the long lifetimes of their excited states, and the ease with which they undergo oxidation and reduction.

The one-electron oxidation products of tris(2,2'-bipyridine)ruthenium(II) and -osmium(II) (ML_3^{3+}) have been known for some time, and the one-electron reduction product of the Ru(II) complex (ML_3^+) has been characterized recently (14, 15, 16, 17, 18). In this article we discuss the reactivities of the luminescent excited state of the +2 oxidation states (*ML_3^{2+}) in terms of the properties of the ground state +2 ion and those of the +1 and +3 ions.

Properties of the Excited States of Tris(2,2'-bipyridine)ruthenium(II) and -osmium(II)

The absorption spectrum of Ru(bpy)$_3^{2+}$ (bpy = 2,2'-bipyridine) is shown in Figure 1. The bands below 200 nm and at 285 nm have been assigned to ligand $\pi \to \pi^*$ transitions by comparisons with the spectrum of protonated bipyridine (19). The two remaining absorption bands at 240 and 450 nm have been assigned to charge-transfer $d \to \pi^*$ transitions (19). The $\pi \to \pi^*$ and $d \to \pi^*$ transitions both result in the formation of the luminescent excited state, and in fact, the luminescent state can be produced by the absorption of light below ~560 nm with close to unit quantum efficiency (20). In the absence of added reagents the excited state decays largely by nonradiative pathways in fluid solution at room temperature, with but a few percent of the excitation energy being released as the emitted red light ($\Phi = 0.042$) (21). Typical excited state lifetimes for Ru(II) and Os(II) polypyridine complexes are given in

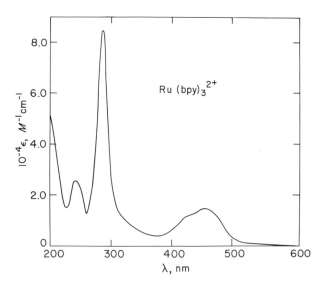

Figure 1. Absorption spectrum of Ru(bpy)$_3^{2+}$ in water at 25°C

Table I. Excited-State Lifetimes for Polypyridineruthenium(II) and -osmium(II) Complexes in Water at 25°C

Ligand, L	RuL_3^{2+} τ_0 (μsec) (32)	OsL_3^{2+} τ_0 (nsec)	$\dfrac{\tau_0(Ru)}{\tau_0(Os)}$
4,4'-(CH$_3$)$_2$bpy	0.33	~ 9	~ 37
bpy	0.60	19	31
bpy-d_8	0.69	32	22
5,6-(CH$_3$)$_2$phen	1.81	63	29
5-(CH$_3$)phen	1.33	69	19
phen	0.92	84	11
5-Cl(phen)	0.94	78	12
5-NO$_2$phen	≤ 0.005	—	—

Table I. It is evident that the osmium lifetimes are shorter and vary more with the nature of the ligands than do the ruthenium lifetimes.

Excited State Spectra. The spectra of *Ru(bpy)$_3^{2+}$ and *Os(bpy)$_3^{2+}$ determined by a laser flash-photolysis technique are presented in Figure 2. (The difference spectrum previously reported (22) for *Ru(bpy)$_3^{2+}$ is in good accord with that used in calculating the spectrum shown in Figure 2.) For *Ru(bpy)$_3^{2+}$ maxima at 360 nm (ϵ 1.3 × 10^4 M^{-1} cm^{-1}) and 430 nm (ϵ 0.6 × 10^4 M^{-1} cm^{-1}) are observed. The spectrum of *Os(bpy)$_3^{2+}$ is very similar with maxima at 360 nm (ϵ 1.8 × 10^4 M^{-1} cm^{-1}) and 460 nm (ϵ 0.5 × 10^4 M^{-1} cm^{-1}). Both are similar to that of the 2,2'-bipyridine anion (23) which is shown in the insert. The radical anion has peaks in this spectral region at 558 and 527 nm (ϵ 0.48 × 10^4 and 0.50 × 10^4 M^{-1} cm^{-1}, respectively), ~ 420 nm (shoulder, ϵ ~ 1.1 × 10^4 M^{-1} cm^{-1}), and at 386 nm (ϵ 2.5 × 10^4 M^{-1} cm^{-1}). The similarity of the spectra of the excited states to that of the bipyridine anion suggests that the observed excited state transitions are ligand localized.

The luminescent state arises from the net transfer of an initially metal-localized d electron to a π^* orbital of the ligand system. Harrigan and Crosby (24) have proposed that the acceptor π^* orbital extends over all three bipyridine ligands. The similarity of the spectra of the excited states and of the bipyridine radical ion is somewhat unexpected if, as proposed, the charge-transfer excited state consists of a Ru(III) center with the additional electron delocalized over the three bipyridine rings (25, 26) rather than as a metal complex of a single bipyridine radical anion. Moreover, since other low energy transitions (for example, bpy → Ru(III) charge-transfer seen at ~ 640 nm in Ru(bpy)$_3^{3+}$, and bpy anion → Ru(III) charge-transfer) might occur in this spectral region, the relative simplicity of the spectra is somewhat surprising. Possibly such transitions are present but either lie outside the spectral range studied or

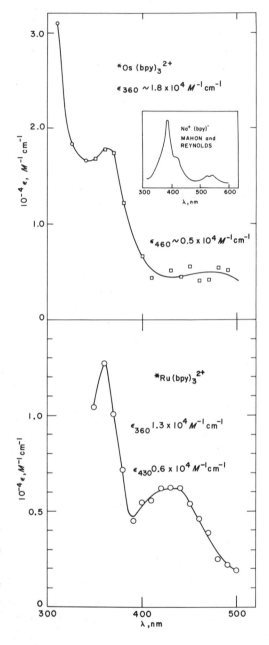

Figure 2. Absorption spectra of the luminescent excited state of $Ru(bpy)_3^{2+}$ (lower spectrum) and $Os(bpy)_3^{2+}$ (upper spectrum) in water at 25°C. The insert shows the spectrum of Na^+bpy^- (23). The errors on the extinction coefficients are estimated to be ±15%.

are obscured by the low resolution and considerable experimental errors associated with these measurements.

Excited State Potentials. In a physical model the formation of the charge-transfer excited state can be regarded as the creation of a separated electron-hole pair. The excited state thus is expected to be both an electron acceptor (hole on the ruthenium center) and an electron donor (excess electron on the bipyridine). A molecular orbital description of the excited state (adapted from Ref. 15) is shown in Figure 3. The excited state has a d^5 metal center characteristic of Ru(bpy)$_3^{3+}$ and an electron in a ligand π^* orbital characteristic of Ru(bpy)$_3^+$. For these reasons, *Ru(bpy)$_3^{2+}$ is expected to exhibit some properties of both Ru(bpy)$_3^{3+}$ and Ru(bpy)$_3^+$.

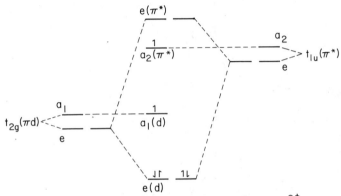

Figure 3. *Molecular orbital diagram for the luminescent excited state of Ru(bpy)$_3^{2+}$. Note that in D_3 microsymmetry the triply degenerate $t_{2g}(\pi)$ and $t_{1u}(\pi^*)$ orbitals are split into two orbitals, one which is doubly degenerate and one which is singly degenerate.*

The above considerations can be put on a more quantitative basis as follows. From absorption and emission spectra the excitation energy of the thermally equilibrated *Ru(bpy)$_3^{2+}$ is 2.10 eV (3, 11). If the entropy differences between ground and excited state molecules are negligible (3, 26), this excitation energy can be considered to be a free energy (Equation 1). For *Os(bpy)$_3^{2+}$, *ΔG has been estimated as 1.78 eV

$$\text{Ru(bpy)}_3^{2+} = {}^*\text{Ru(bpy)}_3^{2+} \qquad {}^*\Delta G = 2.10 \text{ eV} \qquad (1)$$

(11). This excitation free energy can be combined with conventional thermodynamic data (Equations 2 and 3) (15) to give redox potentials for the excited molecule (Equations 4 and 5) (3, 27). Excited state poten-

$$\text{Ru(bpy)}_3^{3+} + e = \text{Ru(bpy)}_3^{2+} \qquad E° = 1.26 \text{ V} \qquad (2)$$

$$\text{Ru(bpy)}_3^{2+} + e = \text{Ru(bpy)}_3^+ \qquad E° = -1.26 \text{ V} \qquad (3)$$

$$\text{Ru(bpy)}_3{}^{3+} + e = {}^*\text{Ru(bpy)}_3{}^{2+} \quad E° = -0.84 \text{ V} \quad (4)$$

$$^*\text{Ru(bpy)}_3{}^{2+} + e = \text{Ru(bpy)}_3{}^{+} \quad E° = +0.84 \text{ V} \quad (5)$$

tials for other $\text{RuL}_3{}^{2+}$ complexes and $\text{OsL}_3{}^{2+}$ complexes are summarized in Table II. For comparison purposes reduction potentials of the ground state complexes are presented in Table III.

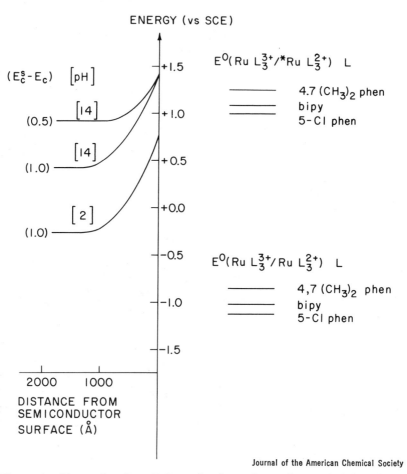

Journal of the American Chemical Society

Figure 4. Energy levels in TiO_2 and polypyridineruthenium(II) complexes (12). The left of the diagram shows the energy of the bottom of the conduction band as a function of pH and distance from the electrode surface for different applied positive potentials.

The right of the diagram shows the energies corresponding to the standard reduction potentials of the Ru(II) complexes. The band bending that occurs at the electrode surface is a consequence of the positive potential applied to the electrode. This band bending constitutes a potential barrier for the transfer of an electron from the bulk of the electrode to a species in solution. This is the condition required for irreversibility of the electron transfer step.

Table II. Excited-State Reduction Potentials for Polypyridine-ruthenium(II) and -osmium(II) Complexes[a] at 25°C

Ligand, L	RuL_3		OsL_3	
	*$E_{3,2}°$, V (32)	*$E_{2,1}°$, V (42)	*$E_{3,2}°$, V (40)	*$E_{2,1}°$, V[b]
4,4'-(CH$_3$)$_2$bpy	−0.94	+0.69	(−1.06)[c]	+0.50
bpy	−0.84	+0.84	−0.96	+0.59
4,7-(CH$_3$)$_2$phen	−1.01	+0.67	—	—
5,6-(CH$_3$)$_2$phen	−0.93	(+0.86)[c]	−1.03	(+0.50)[c]
5-(CH$_3$)phen	−0.90	(+0.89)[c]	−0.99	(+0.54)[c]
phen	−0.87	+0.79	−0.96	(+0.57)[c]
5-Cl(phen)	−0.77	+1.00	−0.85	+0.72

[a] In H$_2$O vs. H$_2$.
[b] Calculated using *$\Delta G/nF$ = 1.78 V for *OsL_3^{2+} and the ground-state potentials given in Table III.
[c] Estimated as described in footnotes g and h of Table III.

Table III. Reduction Potentials (vs. Hydrogen) for Polypyridine-ruthenium and -osmium Complexes at 25°C[a]

Ligand, L	RuL_3		OsL_3	
	$E_{3,2}°$, V (32)	$E_{2,1}°$, V (42)	$E_{3,2}°$, V[d]	$E_{2,1}°$, V
4,4'-(CH$_3$)$_2$bpy	+1.10	−1.37[e]	(+0.72)[h]	−1.31[e]
bpy	+1.26	−1.28	+0.82	−1.22[e]
4,7-(CH$_3$)$_2$phen	+1.09	−1.47	—	—
5,6-(CH$_3$)$_2$phen	+1.20	(−1.34)[i]	+0.75	(−1.28)[h]
5-(CH$_3$)phen	+1.23	(−1.31)[i]	+0.79	(−1.24)[h]
phen	+1.26	−1.36	+0.82	(−1.21)[g]
5-Cl(phen)	+1.36	−1.15	+0.93	−1.09[f]
terpy[a]	+1.25[b]	−1.36[c]	—	—
TPTZ[a]	+1.49[b]	−0.77[c]	—	—

[a] ML_2^{2+}, terpy = 2,2',2''-terpyridine; TPTZ = 2,4,6-tri(2-pyridyl-s-triazine).
[b] From Ref. 15; corrected to H$_2$O is solvent as described in Ref. 32.
[c] From Ref. 15; corrected to H$_2$O as solvent as described in Ref. 27.
[d] In 0.5M H$_2$SO$_4$, 22°C; Brown, G. M., unpublished data.
[e] From Ref. 16; corrected to H$_2$O by subtracting 0.03 V from the value reported for N,N-dimethylformamide.
[f] Creutz, C., unpublished data. Value measured in acetonitrile vs. SCE was corrected to H$_2$O by adding 0.07 V.
[g] Estimated from $E_{3,2}°$ by subtracting 2.54 V which is the average value of ($E_{3,2}°$ − $E_{2,1}°$) for the other RuL_3 entries in this table.
[h] Estimated from ($E_{3,2}°$ − $E_{2,1}°$) = 2.03 V which is the average for this quantity for the other OsL_3 entries in this table.

Excited state potentials also can be obtained from quenching measurements using a series of quenchers of graded potentials (26). In a more direct approach, excited state potentials might be obtained from photocurrent measurements. In practice photocurrents are negligible at metal electrodes since the metal quenches the excited molecules so efficiently (probably by energy transfer) that very little net current can flow (28, 29). Semiconductor electrodes, on the other hand, do not necessarily deactivate the excited state, and recent studies using a biased n-type TiO_2 electrode hold considerable promise for excited state characterization (12). In Figure 4 the energy of the bottom of the conduction band of a TiO_2 electrode is shown as a function of pH and of distance from the electrode surface for different applied positive potentials. The energies of ground-state and excited-state ruthenium RuL_3^{3+}–RuL_3^{2+} couples (Reaction 4, Table II) also are included in Figure 4 for several complexes. At pH = 2 the conduction band edge evidently lies below the energy levels of the excited state couples and therefore anodic photocurrents should be observed for all the complexes at pH = 2. By contrast, at pH = 14 the conduction band edge has moved up to such an extent that electron transfer from the excited state to the semiconductor can no longer take place. Thus no photocurrent should be seen at pH = 14. As shown in Figure 5, pH-dependent sensitized photocurrents of the order of microamps are indeed seen under these conditions when the TiO_2 electrode is connected to a platinum electrode and is illuminated with light in the region 420–550 nm. (Moreover, as is expected, hydrogen evolution occurs at the cathode.) For L = 4,7-$(CH_3)_2$phen, a substantial photocurrent is seen at pH = 5 while for L = 5-Cl(phen), the pH must be lowered to nearly 1 before the photocurrent is observed. This is as expected from the calculated excited-state reduction potentials which differ by 0.24 V. This is an encouraging result since it shows that the potential for the onset of photocurrent might be used to determine

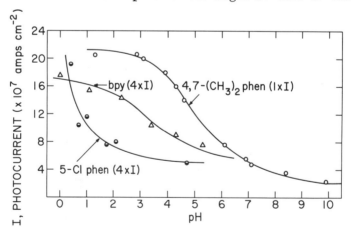

Figure 5. Sensitized photocurrents at TiO_2 vs. pH for the RuL_3^{2+} complexes shown in Figure 4 (12)

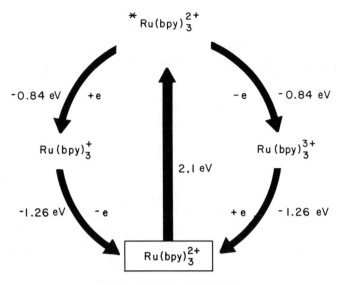

Figure 6. Free energy diagram for the deactivation of the luminescent excited state of $Ru(bpy)_3^{2+}$ to the ground state via $Ru(bpy)_3^+$ or $Ru(bpy)_3^{3+}$

excited-state potentials. Furthermore this approach does not require a luminescent excited state.

Returning to the discussion of excited-state potentials, it is evident from Equation 4 that (in contrast to the ground state, Equation 2) the excited molecule is expected to be a very strong reducing agent. The excited molecule is also a moderately strong oxidant (Equation 5). Note, though, that the excited state is neither as good an oxidant as the +3 ion nor as good a reductant as the +1 ion. These properties are illustrated in Figure 6 which depicts the free energy changes that occur when the excited molecule is deactivated to the ground state by way of $Ru(bpy)_3^+$ or $Ru(bpy)_3^{3+}$. In summary, the excited molecule is expected to undergo either electron loss (at the ligand center) or electron gain (at the metal center) depending on the properties of the other reactant. Both kinds of reaction will be considered in the following sections.

Electron Exchange Reactions. The simplest electron transfer reaction that the excited state can undergo is its self-exchange reaction. This reaction is expected to be rapid. The small value of the Stokes shift (30) for $Ru(bpy)_3^{2+}$ (~ 1.1 kK) indicates that very little distortion occurs on going from the equilibrated ground state to the equilibrated excited state. This interpretation is supported by the very small temperature dependence of the emission spectrum (31). Thus the spectral data indicate that the excited molecule is quite similar in size and shape to the ground state molecule. Since the self-exchange reactions of the ground state molecule are very rapid, the excited-state self exchanges are also likely to be rapid.

Two excited state electron exchange reactions must be considered. The first (Reaction 6) involves the *Ru(bpy)$_3^{2+}$–Ru(bpy)$_3^{3+}$ couple. In

$$*\text{Ru(bpy)}_3^{2+} + \text{Ru(bpy)}_3^{3+} \xrightleftharpoons{*k_{\text{II,III}}} \text{Ru(bpy)}_3^{3+} + *\text{Ru(bpy)}_3^{2+}$$
$$(\pi d)^5(\pi^*) \quad\quad (\pi d)^5 \quad\quad\quad\quad (\pi d)^5 \quad\quad (\pi d)^5(\pi^*)$$

(6)

this exchange reaction the excess π^* electron on the excited molecule is transferred to the vacant π^* orbital on Ru(bpy)$_3^{3+}$. Thus the Ru(III) metal center is largely uninvolved and (at least conceptually) the exchange process is very similar to that of the free bpy–bpy⁻ exchange or the Ru(bpy)$_3^+$–Ru(bpy)$_3^{2+}$ exchange (Reaction 7). Both of the latter

$$\text{Ru(bpy)}_3^+ + \text{Ru(bpy)}_3^{2+} \xrightleftharpoons{k_{\text{I,II}}} \text{Ru(bpy)}_3^{2+} + \text{Ru(bpy)}_3^+$$
$$(\pi d)^6(\pi^*) \quad\quad (\pi d)^6 \quad\quad\quad\quad (\pi d)^6 \quad\quad (\pi d)^6(\pi^*)$$

(7)

reactions are known to be rapid. From electrochemical studies, $k_{\text{I,II}}$ has been estimated as $\geq 10^8$ M⁻¹ sec⁻¹ in N,N-dimethylformamide at 25°C (32, 33). Thus *$k_{\text{II,III}} \sim 1 \times 10^8$ M⁻¹ sec⁻¹ would seem a reasonable guess.

The other excited-state exchange reaction features the Ru(bpy)$_3^+$–*Ru(bpy)$_3^{2+}$ couple (Reaction 8). In this exchange reaction there is

$$\text{Ru(bpy)}_3^+ + *\text{Ru(bpy)}_3^{2+} \xrightleftharpoons{*k_{\text{I,II}}} *\text{Ru(bpy)}_3^{2+} + \text{Ru(bpy)}_3^+$$
$$(\pi d)^6(\pi^*) \quad\quad (\pi d)^5(\pi^*) \quad\quad\quad\quad (\pi d)^5(\pi^*) \quad\quad (\pi d)^6(\pi^*)$$

(8)

no net transfer of the π^* electron, rather a metal d electron is transferred from the Ru(II) core of Ru(bpy)$_3^+$ to the Ru(III) core of *Ru(bpy)$_3^{2+}$. Consequently the exchange process is expected to be very similar to the Ru(bpy)$_3^{2+}$–Ru(bpy)$_3^{3+}$ exchange (Reaction 9) for which Young,

$$\text{Ru(bpy)}_3^{2+} + \text{Ru(bpy)}_3^{3+} \xrightleftharpoons{k_{\text{II,III}}} \text{Ru(bpy)}_3^{3+} + \text{Ru(bpy)}_3^{2+}$$
$$(\pi d)^6 \quad\quad (\pi d)^5 \quad\quad\quad\quad (\pi d)^5 \quad\quad (\pi d)^6$$

(9)

Keene, and Meyer have reported $k_{\text{II,III}} = 1.2 \times 10^9$ M^{-1} sec^{-1} ($\mu = 1.0M$, H$_2$O) at 25°C (34). Thus *$k_{\text{I,II}}$ also is expected to be large and *$k_{\text{I,II}} \sim 1 \times 10^8$ M^{-1} sec^{-1} seems a conservative estimate.

Excited State Reactivity

Ru(bpy)$_3^{2+}$ appears to be inert to photosubstitution at 25°C in water and there is evidence suggesting that the photolabilization observed for this complex at higher temperatures occurs through ligand-field excited states rather than through *Ru(bpy)$_3^{2+}$ (6). In this respect, *Ru(bpy)$_3^{2+}$ resembles Ru(bpy)$_3^{2+}$ and Ru(bpy)$_3^{3+}$, which are also highly inert to substitution. The chemistry of the excited molecule therefore is expected to be restricted predominantly to outer-sphere electron transfer and to energy transfer processes. Three types of bimolecular reaction can thus be envisaged. These are electron loss (Reaction 10), electron addition (Reaction 11), and energy transfer (Reaction 12). We first consider the

$$*\text{RuL}_3^{2+} + Q \rightarrow \text{RuL}_3^{3+} + Q^- \qquad (10)$$

$$*\text{RuL}_3^{2+} + Q \rightarrow \text{RuL}_3^{+} + Q^+ \qquad (11)$$

$$*\text{RuL}_3^{2+} + Q \rightarrow \text{RuL}_3^{2+} + Q^* \qquad (12)$$

electron transfer reactions of excited states in terms of concepts that have proved useful in the interpretation of ground state reactions.

The rates of a large number of ground-state electron transfer reactions have been rationalized in terms of the Marcus equations (35, 36)

$$\begin{aligned} k_{12} &= (k_{11} k_{22} K_{12} f_{12})^{1/2} \\ \log f_{12} &= (\log K_{12})^2 / 4 \log(k_{11} k_{22} / Z^2) \end{aligned} \qquad (13)$$

where k_{11} and k_{22} refer to the exchange reactions and k_{12} and K_{12} to the cross reaction. In terms of this model, the observed electron transfer rates depend on the exchange rates of the couples and the driving force for the electron transfer. On this basis, excited-state reactions are expected to be faster than the corresponding ground-state reactions since the excited-state reactions will have the larger driving forces. In general, the relative magnitudes of the self-exchange rates of the ground- and excited-state couples are more difficult to assess; depending upon the particular system, the rate of exchange of the excited state couple can either be faster or slower than that of the corresponding ground state couple. However, as discussed above, there is good evidence for believing that the ground and excited state self-exchange rates of the polypyridineruthenium(II) and -osmium(II) couples are not too different.

Other differences between the ground- and excited-state couples not allowed for in Equation 13 include spin selection and other symmetry considerations.

Oxidative Quenching. The role of some of the above factors is illustrated in Figure 7 in which rate constants for the quenching of various excited polypyridineruthenium(II) complexes by the potential oxidants Eu_{aq}^{3+}, Cr_{aq}^{3+}, Cu_{aq}^{2+}, and Fe_{aq}^{3+} ions are plotted as a function of the excited state potentials (Equation 4). These aquo ions are all capable of accepting an electron from $*RuL_3^{2+}$ and $*OsL_3^{2+}$. (Reduction potentials of −0.38, −0.40, +0.15, and +0.74 V have been reported for Eu_{aq}^{3+} (37), Cr_{aq}^{3+} (38), Cu_{aq}^{2+} (38), and Fe_{aq}^{3+} (39), respectively.) Quenching is thus expected to occur by either Reaction 10 or Reaction 12. The quenching by the different ions evidently exhibit different free energy dependences with the extreme forms of behavior being shown by the Eu_{aq}^{3+} and Cr_{aq}^{3+} reactions. The relatively large free energy dependence of the Eu_{aq}^{3+} reactions is consistent with an electron transfer quenching

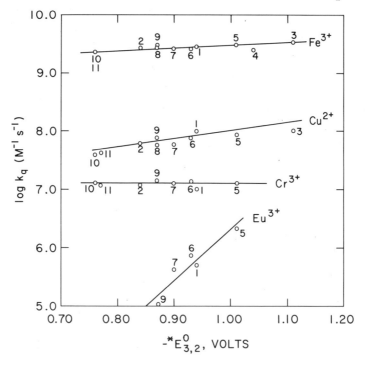

Figure 7. Plot of the logarithm of the rate constant for quenching of $*RuL_3^{2+}$ by Eu_{aq}^{3+}, Cr_{aq}^{3+}, Cu_{aq}^{2+}, and Fe_{aq}^{3+} vs. the $RuL_3^+ - *RuL_3^{2+}$ reduction potential of the RuL_3^{2+} complexes (40): L = 1, 4,4′–(CH₃)₂bpy; 2, bpy; 3, 3,4,7,8-(CH₃)₄phen; 4, 3,5,6,8-(CH₃)₄-phen; 5, 4,7-(CH₃)₂phen; 6, 5,6-(CH₃)₂phen; 7, 5-(CH₃)phen; 8, 5-(C₆H₅)phen; 9, phen; 10, 5-Br(phen); 11, 5-Cl(phen).

Table IV. Rate Constants for Outer-Sphere Reductions by Cr_{aq}^{2+} and Eu_{aq}^{2+}

Oxidant	Cr_{aq}^{2+}	Eu_{aq}^{2+}
	k, M^{-1} sec^{-1}	
$Co(NH_3)_6^{3+}$	0.9×10^{-4} (58)	17×10^{-4} (59)
$Co(en)_3^{3+}$	3.4×10^{-4} (54)	50×10^{-4} (54)
$Co(NH_3)_5py^{3+}$	4.0×10^{-3} (60)	83×10^{-3} (61)
$Co(phen)_3^{3+}$	3.1×10^{1} (51)	90×10^{1} (53)
$Ru(NH_3)_6^{3+}$	2.0×10^{2} (62)	10×10^{2} (63)
$Ru(NH_3)_5py^{3+}$	3.4×10^{3} (64)	54×10^{3} (53)

mechanism while the larger, free-energy independent rates of the Cr_{aq}^{3+} reactions indicate that these reactions proceed by an energy transfer quenching mechanism. This conclusion is consistent with the relative rates of reactions in which both Eu_{aq}^{3+} and Cr_{aq}^{3+} undergo outer-sphere electron transfer (Table IV); it will be seen that the rates of the Eu_{aq}^{3+} reactions are about a factor of twenty faster than the corresponding Cr_{aq}^{3+} reactions (The data given in Table III are for Cr_{aq}^{2+} and Eu_{aq}^{2+} reductions but because of microscopic reversibility and the similarity of the potentials of these two couples, these rate patterns must also hold for the reverse Eu_{aq}^{3+} and Cr_{aq}^{3+} reactions.). This is opposite to the trend in Figure 7. Moreover, energy transfer to Cr_{aq}^{3+} to form the $^4T_{2g}$, ^2E_g, or $^2T_{1g}$ excited states, which have absorption maxima at 17.4 and 15.0 kK, is energetically feasible.

Despite the smaller free energy dependences of their reactions, the quenching by Cu_{aq}^{2+} and Fe_{aq}^{3+} also is ascribed to electron transfer. The very small free energy dependence of the latter reactions arises from the fact that the rates are close to the diffusion-controlled limit. The Cu_{aq}^{2+} reactions pose more of a problem (40). This is further illustrated in Figure 8 in which the rate constants for the oxidation of Cu_{aq}^+ by RuL_3^{3+} complexes (circles) are compared with the rates of reaction of the *RuL_3^{2+} complexes with Cu_{aq}^{2+} (squares). The smaller free energy dependence of the quenching reactions is again evident. One possible explanation for the small free energy dependence of the Cu_{aq}^{2+} quenching reactions is that these reactions proceed in part by an energy transfer pathway. To examine this possibility, the rates of reaction of Cu_{aq}^{2+} with RuL_3^+, generated by pulse radiolysis, were measured. These reactions are included in Figure 8 (squares). It will be seen that these reactions, which definitely involve electron transfer, also exhibit small free energy dependences and indeed that their rates are consistent with the rates for reaction of Cu_{aq}^{2+} with *RuL_3^{2+}. This observation supports an electron transfer mechanism for the Cu_{aq}^{2+} quenching reactions. It is of interest that the reactions of *RuL_3^{2+} and RuL_3^+ with Cu_{aq}^{2+} have in

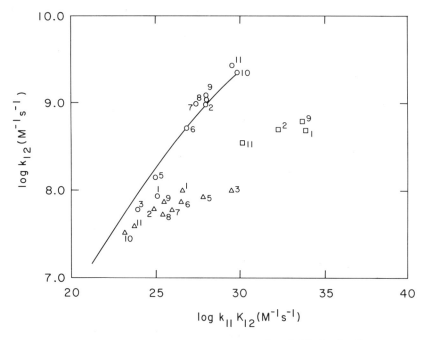

Figure 8. Plot of the logarithm of the rate constant for electron transfer vs. the logarithm of the product of the exchange rate constant of the ruthenium complex and the equilibrium constant for the reaction: (circles) Cu_{aq}^+ + RuL_3^{3+}; (triangles) $*RuL_3^{2+}$ + Cu_{aq}^{2+}; (squares) RuL_3^+ + Cu_{aq}^{2+}; ruthenium complexes numbered as in Figure 7 (40).

common at least one feature not shared by the RuL_3^{3+} reactions; namely, the transfer of a π^* electron in the oxidation–reduction step. This could be responsible for the different free energy dependences of the RuL_3^+, $*RuL_3^{2+}$, and RuL_3^{3+} reactions.

Equation 13 can be used to calculate rate constant ratios for the ground- and excited-state reactions. Applied to the Eu(III) reactions, the ratio of the rate constants for the reactions

$$*RuL_3^{2+} + Eu_{aq}^{3+} \rightarrow RuL_3^{3+} + Eu_{aq}^{2+} \tag{14}$$

$$RuL_3^{2+} + Eu_{aq}^{3+} \rightarrow RuL_3^{3+} + Eu_{aq}^{2+} \tag{15}$$

(L is 4,7-$(CH_3)_2$phen), is calculated to be 10^{23}, in excellent agreement with the observed ratio of 3×10^{23} (13). This suggests that, at least for this particular system, the relative reactivity of the excited and ground states is primarily controlled by thermodynamic factors. It is also interesting to compare rate constants for oxidation of $*Ru(bpy)_3^{2+}$ with those for oxidation of $Ru(bpy)_3^+$. As mentioned above, the self-exchange rates

Table V. Rate Constants for Reduction by
$*Ru(bpy)_3^{2+}$ and $Ru(bpy)_3^+$ at 25°C

Oxidant	$*Ru(bpy)_3^{2+}$	$Ru(bpy)_3^+$
	k, M^{-1} sec^{-1}	
Cu_{aq}^{2+}	6.6×10^7 (40)	5.2×10^8 (40)
O_2	3.3×10^9 (11)	4.0×10^9 (42)
$Ru(NH_3)_6^{3+}$	3.1×10^9 (3)	4.7×10^9 (42)
Eu_{aq}^{3+}	2.1×10^5 (32)	5.7×10^7 (17)

for the $*Ru(bpy)_3^{2+}$–$Ru(bpy)_3^{3+}$ and the $Ru(bpy)_3^+$–$Ru(bpy)_3^{2+}$ couples are expected to be large and similar since both exchange processes involve transfer of a π^* electron. Since $Ru(bpy)_3^+$ is a stronger reductant than $*Ru(bpy)_3^{2+}$, rate constants for oxidation of $Ru(bpy)_3^+$ should be greater than for oxidation of $*Ru(bpy)_3^{2+}$ (from Equation 13 without f corrections, the rate ratio should be $\sim 3 \times 10^3$). In Table V it is evident that for the four oxidants which have been used with both reductants, the $Ru(bpy)_3^+$ reactions are, as predicted, more rapid than the $*Ru(bpy)_3^{2+}$ reactions. Only with Eu_{aq}^{3+} and Cu_{aq}^{2+} as oxidants is the reactivity difference substantial since the other rate constants are very close to diffusion controlled.

Reductive Quenching and the Behavior of $Ru(bpy)_3^+$. The first evidence for the reduction of $*Ru(bpy)_3^{2+}$ to $Ru(bpy)_3^+$ (Reaction 11) came from kinetic studies (27, 41). The excited-state emission was found to be quenched efficiently by various inorganic reducing agents and the quenching rate constants measured were in rough accord with the predictions of the Marcus theory. Some of the same reductants, namely $Fe(CN)_6^{4-}$, $Ru(NH_3)_6^{2+}$, and Eu_{aq}^{2+}, have been used to probe the relative reactivities of $*Ru(bpy)_3^{2+}$ and $*Os(5\text{-}Clphen)_3^{2+}$, whose excited state reduction potentials are $+0.84$ and $+0.70$ V, respectively. In Table VI it is evident that the quenching rate constants are slower for the osmium complex than for the ruthenium complex. This is as expected for quenching according to Reaction 11 since the osmium excited state is a poorer oxidant than the ruthenium excited state.

Table VI. Rate Constants for Reaction of $*ML_3^{2+}$ with Inorganic Reductants at 25°C at 0.5 to 1.0M Ionic Strength

Quencher	$Ru(bpy)_3^{2+}$	$Os(5\text{-}Clphen)_3^{2+}$
	k, M^{-1} sec^{-1}	
$Fe(CN)_6^{4-}$	3.5×10^9 (27)	1×10^8
$Ru(NH_3)_6^{2+}$	2.4×10^9 (27)	1×10^9
Eu_{aq}^{2+}	2.8×10^7 (27)	$\sim 5 \times 10^6$

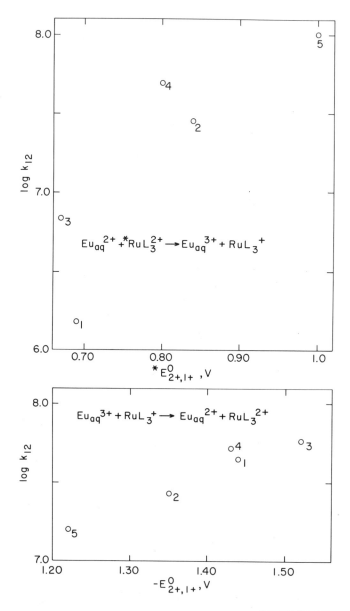

Figure 9. (Top) Plot of the logarithm of the rate constant for the quenching of $^*RuL_3^{2+}$ by Eu_{aq}^{2+} vs. the $^*RuL_3^{2+}$–RuL_3^+ reduction potential (42); (bottom) plot of the logarithm of the rate constant for the oxidation of RuL_3^+ by Eu_{aq}^{3+} vs. RuL_3^{2+}–RuL_3^+ reduction potential in CH_3CN vs. SCE: L = 1, 4,4'-$(CH_3)_2$bpy; 2, bpy; 3, 4,7-$(CH_3)_2$phen; 4, phen; 5, 5-Cl(phen).

The most direct evidence for reduction of *Ru(bpy)$_3^{2+}$ is provided by flash-photolysis experiments (*17*). Aqueous solutions containing Eu$_{aq}^{2+}$ and Ru(bpy)$_3^{2+}$ were excited using a frequency-doubled neodymium laser (530 nm, ~ 25 nsec pulse). A 500-nm absorbing transient was observed to form with the same rate as *Ru(bpy)$_3^{2+}$ decay. The latter was determined independently by monitoring the time dependence of the emission from the solution. On a longer time scale the transient decayed, with the absorbance of the solution returning to its preflash absorbance. These observations are consistent with the following scheme (Reactions 16, 17, and 18).

$$\text{Ru(bpy)}_3^{2+} \xrightarrow{h\nu} {}^*\text{Ru(bpy)}_3^{2+} \qquad (16)$$

$$^*\text{Ru(bpy)}_3^{2+} + \text{Eu}_{aq}^{2+} \xrightarrow{k_q} \text{Ru(bpy)}_3^+ + \text{Eu}_{aq}^{3+} \qquad (17)$$

$$\text{Ru(bpy)}_3^+ + \text{Eu}_{aq}^{3+} \xrightarrow{k_t} \text{Ru(bpy)}_3^{2+} + \text{Eu}_{aq}^{2+} \qquad (18)$$

Similar observations have been made with other polypyridineruthenium(II) complexes and europium(II) (*42*). In Figure 9 the logarithm of the quenching-rate constant (top) is plotted against the excited-state reduction potential. The Eu$_{aq}^{2+}$ quenching rate constants range from ~ 10^6 to 10^8 M^{-1} sec^{-1} while the excited state potentials range from 0.7 to 1.0 V. As expected from the Marcus theory, the quenching rate constants increase as the oxidizing power of the excited molecule increases. The rate constants for the "back reactions" (oxidation of RuL$_3^+$ by Eu$_{aq}^{3+}$) also increase with driving force, as shown in the lower portion of Figure 9. Formation of Ru(bpy)$_3^+$ by reductive quenching of *Ru(bpy)$_3^{2+}$ using organic donors in the solvents methanol (*43*) and acetonitrile (*18*) also has been accomplished. The flash photolysis observations indicate that the spectrum of Ru(bpy)$_3^+$ does not shift greatly with solvent. Further, the quenching rate patterns are consistent with the calculated *Ru(bpy)$_3^{2+}$–Ru(bpy)$_3^+$ potential of +0.84 V.

Comparison of *Ru(bpy)$_3^{2+}$ Oxidation and Reduction Reactions

*Ru(bpy)$_3^{2+}$ can either give up or take on an electron depending on circumstances. This idea is illustrated for two reagent pairs in Reactions 19, 20, 21, and 22.

$$^*\text{Ru(bpy)}_3^{2+} + \text{Fe}_{aq}^{3+} \underset{}{\overset{K = 6 \times 10^{26}}{\rightleftharpoons}} \text{Ru(bpy)}_3^{3+} + \text{Fe}_{aq}^{2+} \qquad (19)$$

$$^*\text{Ru(bpy)}_3^{2+} + \text{Fe}_{aq}^{2+} \underset{K_{ox}/K_{red} \sim 10^{25}}{\overset{K = 50}{\rightleftharpoons}} \text{Ru(bpy)}_3^+ + \text{Fe}_{aq}^{3+} \quad (20)$$

$$^*\text{Ru(bpy)}_3^{2+} + \text{Eu}_{aq}^{3+} \overset{K \sim 10^7}{\rightleftharpoons} \text{Ru(bpy)}_3^{3+} + \text{Eu}_{aq}^{2+} \quad (21)$$

$$^*\text{Ru(bpy)}_3^{2+} + \text{Eu}_{aq}^{2+} \underset{K_{ox}/K_{red} \sim 10^{-15}}{\overset{K \sim 10^{22}}{\rightleftharpoons}} \text{Ru(bpy)}_3^+ + \text{Eu}_{aq}^{3+} \quad (22)$$

The equilibrium constants given above the arrows were calculated from the Fe_{aq}^{3+}–Fe_{aq}^{2+} and Eu_{aq}^{3+}–Eu_{aq}^{2+} potentials and the excited-state potentials given earlier. In several systems it has been possible to obtain rate constants for both oxidative and reductive quenching by the same couple. We can explore how the ratio of the rate constants for the two processes (i.e. $k_{ox}/k_{red} = k_{19}/k_{20}$ or k_{21}/k_{22}) depends on the relative driving forces for the two processes. For purposes of this comparison we shall assume that both excited state self-exchange processes ($^*\text{Ru(bpy)}_3^{2+}$–Ru(bpy)_3^{3+} and Ru(bpy)_3^+–$^*\text{Ru(bpy)}_3^{2+}$) have the same rate constant. With this

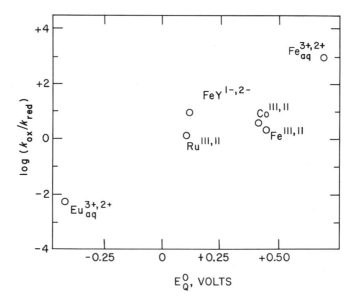

Figure 10. Plot of the logarithm of the ratio of the rate constant for quenching by the oxidized form of a redox couple to that for quenching by the reduced form vs. the reduction potential of the redox couple (27, 50): the couples are $\text{Eu}_{aq}^{3+,2+}$, $\text{Ru(NH}_3)_6^{3+,2+}$, $\text{Fe(EDTA)}^{1-,2-}$ (50), $\text{Co(phen)}_3^{3+,2+}$, $\text{Fe(CN)}_6^{3-,4-}$, and $\text{Fe}_{aq}^{3+,2+}$

assumption and neglect of the Marcus f term we then have, from Equation 13

$$\frac{k_{ox}}{k_{red}} = \left(\frac{K_{ox}}{K_{red}}\right)^{1/2} \tag{23}$$

which can be expressed in terms of the reduction potential of the quenching couple ($E_Q°$) and the excited state reduction potentials (*$E_{3,2}°$ and *$E_{2,1}°$) as Equation 24.

$$\log (k_{ox}/k_{red}) = \frac{16.9}{2} (2 E_Q° - {}^*E_{3,2}° - {}^*E_{2,1}°) \tag{24}$$

This equation suggests that a plot of $\log(k_{ox}/k_{red})$ vs. the reduction potential of the quencher should be linear with an intercept of zero for *Ru(bpy)$_3^{2+}$ if the estimated excited-state reduction potentials are correct. This prediction is borne out in Figure 10. Although the slope of the plot is smaller than predicted by Equation 24 (for most of the quenchers the f corrections are not negligible), the predicted behavior is observed. The mode of data presentation given in Figure 7 emphasizes the amphoteric character of the excited state. The excited molecule can either take on or give up an electron. Which process dominates is determined by kinetic and thermodynamic considerations, and when both processes have comparable driving forces, the kind of reaction occurring largely will be determined by the relative concentrations of the reduced and oxidized forms of the quencher.

In Figures 11, 12, and 13, data for reductions by Fe_{aq}^{2+}, Cu_{aq}^+, and Eu_{aq}^{2+} are summarized. For many of the data points, the reaction studied was an oxidation of the aquo-ion, and for these systems the rate constant for the reverse reaction was calculated from the equilibrium constant K_{12}. Because the oxidants are diverse in properties, an attempt to normalize the data has been made by plotting as ordinate $\log K_{12}k_{22}$ where k_{22} is the self-exchange rate of the oxidant couple. It is obvious that the data for *Ru(bpy)$_3^{2+}$ reactions (solid triangles, $k_{22} \sim 10^8$ M^{-1} sec^{-1}) fit smoothly with those for reductions of ground-state oxidants. These comparisons suggest that there is nothing abnormal about the excited-state reactions once allowance for their large free energy changes is made. Remarkably, these comparisons also imply that the self-exchange rates for the $Fe_{aq}^{2+,3+}$, $Cu_{aq}^{+,2+}$ and $Eu_{aq}^{2+,3+}$ couples are very similar, or at least, that the cross-reaction rates (once corrected for free energy differences) are not particularly sensitive to whether the reductant is a Fe_{aq}^{2+}, Cu_{aq}^+, or Eu_{aq}^{2+} ion.

Quenching by Polypyridine Complexes. Polypyridineruthenium(III) complexes are very strong oxidants and have been shown to oxidize hydrox-

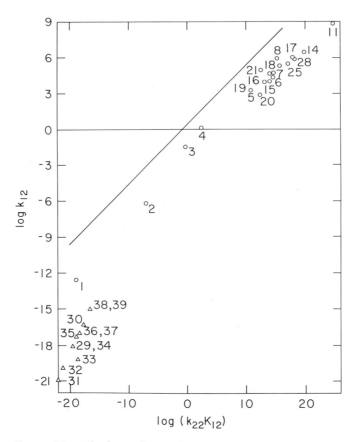

Figure 11. The logarithm of the rate constant for the oxidation of Fe_{aq}^{2+} vs. the logarithm of the product of the equilibrium constant and the oxidant self-exchange rate. For points 1 through 4 the oxidants are V_{aq}^{3+} (45), $Ru(NH_3)_6^{3+}$ (46), $Ru(NH_3)_5(isonicotinamide)^{3+}$ (47), and cis-$Ru(NH_3)_4(bpy)^{3+}$ (47), respectively. For points 5 to 9 the oxidants are $Os(bpy)_3^{3+}$ (48), $Fe(bpy)_3^{3+}$ (49), $Fe(phen)_3^{3+}$ (49), $Ru(bpy)_3^{3+}$ (48), and $Co(phen)_3^{3+}$ (51), respectively. For points 11, 14, and 15 the oxidants are $Cr(bpy)_3^{3+}$ (52), $Ru(5$-$Clphen)_3^{3+}$ (32), and $Ru[3,4,7,8$-$(CH_3)_4phen]_3^{3+}$ (32), respectively. For points 16 to 20 the oxidant is FeL_3^{3+} (49) with L = 16, 5-$(CH_3)phen$; 17, 5-$(NO_2)phen$; 18, 5-$Cl(phen)$; 19, 5,6-$(CH_3)_2phen$; 20, 4,4'-$(CH_3)_2bpy$. For point 21 the oxidant is $Fe(terpy)_2^{3+}$ (49). For points 25 to 28 the oxidant is RuL_3^{3+} (32) with L = 25, 5,6-$(CH_3)_2phen$, 26, 5-$(CH_3)phen$; 27, 5-$(C_6H_5)phen$; 28, phen. Self-exchange rates and reduction potentials for the above oxidants are summarized in Ref. 53 and given in the original references. For points 29 to 39 the reaction is $Fe_{aq}^{2+} + RuL_3^{3+} = Fe_{aq}^{3+} + *RuL_3^{2+}$ (32) with L = 29, 4,4'-$(CH_3)_2bpy$; 30, bpy; 31, 3,4,7,8-$(CH_3)_4phen$; 32, 3,5,6,8-$(CH_3)_4phen$; 33, 4,7-$(CH_3)_2phen$; 34, 5,6-$(CH_3)_2phen$; 35, 5-$(CH_3)phen$; 36, 5-$(C_6H_5)phen$; 37, phen; 38, 5-$Br(phen)$; 39, 5-$Cl(phen)$. Excited-state reduction potentials are given in Ref. 32. The self-exchange rate used for the $*RuL_3^{2+}$-RuL_3^{3+} couple was 1.0×10^8 M^{-1} sec^{-1}. The line shown is that calculated from Equation 13, assuming f = 1 and $k_{11} = 4.2$ M^{-1} sec^{-1}.

ide ion to dioxygen (9). The RuL_3^+ species are, by contrast, very strong reductants and should be capable of reducing hydrogen ion to dihydrogen. Thus for solar energy conversion and storage by the photodecomposition of water, it would be desirable to generate both RuL_3^{3+} and RuL_3^+ from $*RuL_3^{2+}$.

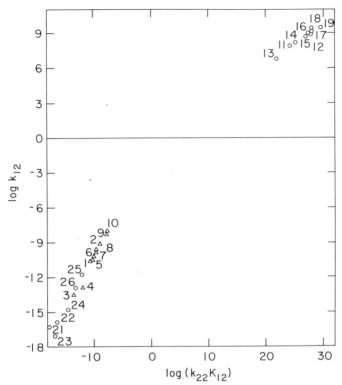

Figure 12. The logarithm of the rate constant for the oxidation of Cu_{aq}^+ vs. the logarithm of the product of the equilibrium constant and the oxidant self-exchange rate.

For points 1–10 (triangles) the reaction is $Cu_{aq}^+ + RuL_3^{3+} = Cu_{aq}^{2+} + *RuL_3^{2+}$ with L = 1, 4,4'-(CH$_3$)$_2$bpy; 2, bpy; 3, 3,4,7,8-(CH$_3$)$_4$-phen; 4, 4,7-(CH$_3$)$_2$phen; 5, 5,6-(CH$_3$)$_2$phen; 6, 5-(CH$_3$)phen; 7, 5-(C$_6$H$_5$)phen; 8, phen; 9, 5-Cl(phen); 10, 5-Br(phen). The value of k_{22} used for the $*RuL_3^{2+} - RuL_3^{3+}$ couple was 1×10^8 M^{-1} sec^{-1}. For points 11 to 24 the reaction is $Cu_{aq}^+ + RuL_3^{3+} = Cu_{aq}^{2+} + RuL_3^{2+}$ with L = 11, 4,4'-(CH$_3$)$_2$bpy; 12, bpy; 13, 3,4,7,8-(CH$_3$)$_4$-phen; 14, 4,7-(CH$_3$)$_2$phen; 15, 5,6-(CH$_3$)$_2$phen; 16, 5-(CH$_3$)phen; 17, 5-(C$_6$H$_5$)phen; 18, phen; 19, 5-Cl(phen); 20, 5-Br(phen). The self-exchange rate k_{22} used for the $RuL_3^{2+}–RuL_3^{3+}$ couple was 2×10^9 M^{-1} sec^{-1}. For points 21 to 24, the reaction is $Cu_{aq}^+ + RuL_3^{2+} \rightarrow Cu_{aq}^{2+} + RuL_3^+$ with L = 21, 4,4'-(CH$_3$)$_2$bpy; 22, bpy; 23, phen; 24, 5-Cl(phen). The self-exchange rate k_{22} used was 1×10^8 M^{-1} sec^{-1}. For points 41 and 42 the reaction was $Cu_{aq}^+ + OsL_3^{3+} \rightarrow Cu_{aq}^{2+} + *OsL_3^{2+}$ with L = 41, bpy; 42, 5,6-(CH$_3$)$_2$phen. The self-exchange rate used for the $*OsL_3^{2+}–OsL_3^{3+}$ couple was 1×10^8 M^{-1} sec^{-1}. The data are taken from Ref. 40.

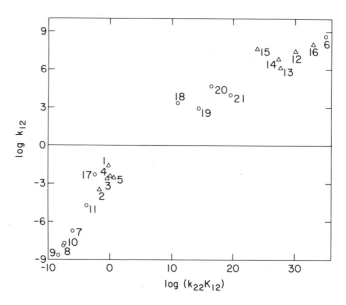

Figure 13. The logarithm of the rate constant for the oxidation of Eu_{aq}^{2+} vs. the logarithm of the product of the equilibrium constant and the oxidant self-exchange rate.
For points 1 to 5 the reaction is $Eu_{aq}^{2+} + RuL_3^{3+} = Eu_{aq}^{3+} + $*$RuL_3^{2+}$ (32) with L = 1, 4,4'-$(CH_3)_2$bpy; 2, 4,7-$(CH_3)_2$phen; 3, 5,6-$(CH_3)_2$phen; 4, 5-(CH_3)phen; 5, phen and $k_{22} = 1 \times 10^9$ M^{-1} sec^{-1}. For point 6 the reaction is $Eu_{aq}^{2+} + Ru[4,7-(CH_3)_2phen]_3^{3+} = Eu_{aq}^{3+} + Ru[4,7-(CH_3)_2phen]_3^{2+}$ ($k_{22} = 2 \times 10^9$ M^{-1} sec^{-1}). For points 7 to 11 the reaction is $Eu_{aq}^{2+} + RuL_3^{2+} = Eu_{aq}^{3+} + RuL_3^{+}$ (42) with L = 7, bpy; 8, 4,4'-$(CH_3)_2$bpy; 9, 4,7-$(CH_3)_2$phen; 10, phen; 11, 5-Cl(phen) (k_{22} used was 1×10^8 M^{-1} sec^{-1}). For points 12 to 16 the reaction was $Eu_{aq}^{2+} + $*$RuL_3^{2+} = Eu_{aq}^{3+} + RuL_3^{+}$ (42) (the value of k_{22} used was 1×10^8 M^{-1} sec^{-1}) with L = 12, bpy; 13, 4,4'-$(CH_3)_2$bpy; 14, 4,7-$(CH_3)_2$phen; 15, phen; 16, 5-Cl(phen). For points 17 to 21 the oxidants are $Co(en)_3^{3+}$ (54), $Ru(NH_3)_6^{3+}$ (55), $Co(phen)_3^{3+}$ (53), $Ru(NH_3)_5py^{3+}$ (53), Fe_{aq}^{3+} (56). Self-exchange rates and reduction potentials for these oxidants are summarized in (53).

Reaction of two excited state molecules according to Reaction 25a is

$$2 \,{}^*RuL_3^{2+} \rightarrow RuL_3^{3+} + RuL_3^{+} \qquad (25a)$$

$$2 \,{}^*\Delta G/nF > (E_{3,2}° - E_{2,1}°) \qquad (25b)$$

thermodynamically favored if Equation 25b is satisfied, a condition which is certainly met for all the ruthenium and osmium complexes. For *$Ru(bpy)_3^{2+}$, evidence for excited state disproportionation has been sought from both absorbance and emission measurements but Reaction 25a is evidently slow with $k \leqslant 10^7$ M^{-1} sec^{-1} (44). The disproportionation can, however, be catalyzed by a pair of redox quenchers, e.g. $Ru(NH_3)_6^{2+}$

and $Ru(NH_3)_6^{3+}$; the reductant generates $Ru(bpy)_3^+$ from the excited state while the oxidant generates $Ru(bpy)_3^{3+}$. (This particular pair of quenchers would not give high yields of the disproportionation products since the back-reaction rate constants are diffusion controlled.)

The reaction of one excited molecule with its parent ground-state molecule (Reaction 26a) proceeds spontaneously only when condition 26b

$$*ML_3^{2+} + ML_3^{2+} \rightarrow ML_3^{3+} + ML_3^+ \qquad (26a)$$

$$*\Delta G/nF \geqslant (E_{3,2}^° - E_{2,1}^°) \qquad (26b)$$

is satisfied. Inspection of Table III indicates that $(E_{3,2}^°-E_{2,1}^°)$ is about 2.5 V for RuL_3^{2+} complexes and about 2.0 V for the OsL_3^{2+} complexes. These potential differences substantially exceed the excitation free energies of $\sim$ 2.1 and $\sim$ 1.8 V for RuL_3^{2+} and OsL_3^{2+}, respectively; thus, Equation 26b is not satisfied for these systems. (In fact, the reverse of Reaction 26a occurs readily and has been observed in electrochemical experiments from its striking chemiluminescence in acetonitrile (15) and in water (56). The reaction of $*Ru(bpy)_3^{2+}$ with ML_3^{3+} also is very rapid (44). A light-induced disproportionation analogous to Reaction 26a however is feasible if the excited and ground state molecules are different (Reaction 27).

$$*ML_3^{2+} + M'L_3'^{2+} \rightleftarrows \begin{array}{l} ML_3^{3+} + M'L_3'^+ \quad (27a) \\ ML_3^+ + M'L_3'^{3+} \quad (27b) \end{array}$$

Reaction 27a can occur when $*E_{3,2}^° < (E_{2,1}^°)'$ while Reaction 27b is favored when $*E_{2,1}^° > (E_{3,2}^°)'$.

A complicating but interesting feature of reactions between the metal(II) polypyridine complexes is that energy transfer processes can occur in competition with electron transfer processes. Energy transfer certainly must be taken into account when M and M' are identical ($\Delta G \sim 0$), while, when M is Ru but M' is Os, exergonic energy transfer can occur. On the other hand, energy transfer from $*OsL_3^{2+}$ to RuL'_3^{2+} should be uphill and is expected to be slow. These predictions are borne out by the quenching rate constants for the following reactions (in all of which electron transfer Reactions 27a and 27b are thermodynamically unfavorable).

$$*Ru(4,7\text{-}(CH_3)_2phen)_3^{2+} + Os(bpy)_3^{2+}$$
$$\xrightarrow[\substack{K \geqslant 10^5 \\ k_q = 1.9 \times 10^9 \text{ M}^{-1} \text{ sec}^{-1}}]{} Ru(4,7\text{-}(CH_3)_2phen)_3^{2+} + *Os(bpy)_3^{2+} \qquad (28)$$

$$*\text{Ru(bpy)}_3^{2+} + \text{Ru(terpy)}_2^{2+}$$

$$\xrightarrow[\substack{K \sim 1 \\ k_q = 1.5 \times 10^9 \text{ M}^{-1} \text{ sec}^{-1}}]{} \text{Ru(bpy)}_3^{2+} + *\text{Ru(terpy)}_2^{2+} \quad (29)$$

$$*\text{Os(5-Clphen)}_3^{2+} + \text{Ru(terpy)}_2^{2+}$$

$$\xrightarrow[\substack{K \sim 10^{-5} \\ k_q \leqslant 1 \times 10^8 \text{ M}^{-1} \text{ sec}^{-1}}]{} \text{Os(5-Clphen)}_3^{2+} + *\text{Ru(terpy)}_2^{2+} \quad (30)$$

(It is worth mentioning that Fe(phen)_3^{2+} quenching of both $*\text{Ru(bpy)}_3^{2+}$ and $*\text{Os(5-Clphen)}_3^{2+}$ proceeds with $k_q > 2 \times 10^9$ M^{-1} sec^{-1}. As redox quenching is uphill in both cases, an energy transfer quenching process yielding $*\text{Fe(phen)}_3^{2+}$ seems likely. The results suggest that the excitation energy of the acceptor Fe(II) state lies below that of both $*\text{RuL}_3^{2+}$ and $*\text{OsL}_3^{2+}$ so that $*E < 1.8$ V for $*\text{FeL}_3^{2+}$. Additional evidence for this (or perhaps a different) FeL_3^{2+} excited state comes from flash-photolysis experiments in which transient bleaching of the FeL_3^{2+} ground state absorption (L = bpy, d^8-bpy, 4,4'-(CH$_3$)$_2$bpy, phen, 5-Clphen, 4,7-(CH$_3$)$_2$phen) was observed. From these experiments the lifetimes of the FeL_3^{2+} excited states in water at 25°C are estimated to lie in the range 0.1–2 nsec.)

Because energy transfer quenching of $*\text{OsL}_3^{2+}$ by ruthenium(II)-polypyridine complexes is expected to be slow (that is, considerably slower than diffusion controlled, cf Reaction 30), it seems reasonable that quenching of $*\text{Os(5-Clphen)}_3^{2+}$ by Ru(TPTZ)_2^{2+} for which k_q is 2.6×10^9 M^{-1} sec^{-1} does occur by oxidation of the excited state as shown in Reaction 31.

$$*\text{Os(5-Clphen)}_3^{2+} + \text{Ru(TPTZ)}_2^{2+}$$

$$\xrightarrow[K \sim 10]{} \text{Os(5-Clphen)}_3^{3+} + \text{Ru(TPTZ)}_2^{+} \quad (31)$$

Further studies of the above and related systems are being made to confirm these preliminary conclusions, but in any case it is clear that Reactions 27a or 27b must occur for certain pairs of complexes. When these electron transfer reactions are thermodynamically favored, they should be nearly diffusion controlled because the self-exchange rates for all the ground- and excited-state polypyridine couples are very high. Energy conversion (and temporary storage) according to Reaction 27 is highly attractive because the driving force for the excited-state capture is small so that little of the excited-state free energy is lost on conversion to the electron transfer products. With this energy conversion scheme (as with most others involving solution reactions of excited molecules), problems in long term energy storage arise because of "back reactions;" the newly formed products (for example ML_3^{3+} and $M'L'_3^+$) can undergo very rapid reaction, leading to regeneration of the species present before irradiation and the production of heat. Schemes devised for the defeat of back reactions include the use of semiconductor electrodes, membranes, micelles, and the exploitation of the inverted region (*44*). Some of these should be applicable to the present problem. (Reactions analogous to Reaction 27 may be responsible for the observations reported by Sprintschnik, Sprintschnik, Kirsch, and Whitten (*13*). These authors initially reported the formation of hydrogen and oxygen from the photodecomposition of water mediated by monolayers of RuL_3^{2+} complexes but were unable to reproduce these observations using purified materials. The presence of RuL'_3^{2+} impurities in the original sample could have given rise to +1 and +3 complexes which might have produced hydrogen and oxygen in subsequent reactions.) The important point is that reactions such as Reaction 27 convert one ground-state and one excited-state molecule into a powerful oxidant and a powerful reductant, each of which lives longer than the excited state and which can, as a consequence, undergo a larger variety of reactions.

In this chapter, we have attempted to describe the luminescent charge-transfer excited state of $Ru(bpy)_3^{2+}$ as a novel oxidation state of a ruthenium complex. Its lifetime is so long that it equilibrates with the surrounding solution, and consequently, some of its thermodynamic properties (namely its oxidation–reduction potentials) can be calculated. In fact, it is even possible to determine its electrochemical properties directly. With regard to substitution reactions, $^*Ru(bpy)_3^{2+}$ is inert like $Ru(bpy)_3^{2+}$ and $Ru(bpy)_3^{3+}$. In its electron-transfer reactions, its reactivity can be related to that of $Ru(bpy)_3^+$ and $Ru(bpy)_3^{3+}$. In summary, the physical and chemical properties of the excited state are presently as well characterized as those of many ground state complexes.

Acknowledgment

This research was carried out at Brookhaven National Laboratory under contract with the U. S. Department of Energy and was supported by its Division of Basic Energy Sciences.

Literature Cited

1. Gafney, H. D., Adamson, A. W., *J. Am. Chem. Soc.* (1972) **94**, 8238.
2. Bock, C. R., Meyer, T. J., Whitten, D. G., *J. Am. Chem. Soc.* (1974) **96**, 4710.
3. Navon, G., Sutin, N., *Inorg. Chem.* (1974) **13**, 2159.
4. Laurence, G. A., Balzani, V., *Inorg. Chem.* (1974) **13**, 2976.
5. Balzani, V., Moggi, L., Manfrin, M. F., Bolletta, F., Laurence, G. S., *Coord. Chem. Rev.* (1975) **15**, 321.

6. Van Houten, J., Watts, R. J., *J. Am. Chem. Soc.* (1976) **98**, 4853.
7. Winterle, J. S., Kliger, D. S., Hammond, G. S., *J. Am. Chem. Soc.* (1976) **98**, 3719.
8. Demas, J. N., Harris, E. W., McBride, E. P., *J. Am. Chem. Soc.* (1977) **99**, 3547.
9. Creutz, C., Sutin, N., *Proc. Nat. Acad. Sci., U.S.A.* (1975) **72**, 2858.
10. Young, R. C., Meyer, T. J., Whitten, D. G., *J. Am. Chem. Soc.* (1975) **97**, 4781.
11. Lin, C.-T., Sutin, N., *J. Phys. Chem.* (1976) **80**, 97.
12. Clark, W. D. K., Sutin, N., *J. Am. Chem. Soc.* (1977) **99**, 4676.
13. Sprintschnik, G., Sprintschnik, H. W., Kirsch, P. P., Whitten, D. G., *J. Am. Chem. Soc.* (1977) **99**, 9747.
14. Baxendale, J. H., Fiti, M., *J. Chem. Soc., Dalton Trans.* (1972) 1995.
15. Tokel-Takvoryan, N. E., Hemingway, R. E., Bard, A. J., *J. Am. Chem. Soc.* (1973) **95**, 6582.
16. Saji, T., Aoyagui, S., *J. Electroanal. Chem. Interfacial Electrochem.* (1975) **58**, 401.
17. Creutz, C., Sutin, N., *J. Am. Chem. Soc.* (1976) **98**, 6384.
18. Anderson, C. P., Salmon, D. J., Meyer, T. J., Young, R. J., *J. Am. Chem. Soc.* (1977) **99**, 1980.
19. Lytle, F. E., Hercules, D. M., *J. Am. Chem. Soc.* (1969) **91**, 253.
20. Demas, J. N., Crosby, G. A., *J. Am. Chem. Soc.* (1971) **93**, 2841.
21. Van Houten, J., Watts, R. J., *J. Am. Chem. Soc.* (1975) **97**, 3843.
22. Bensasson, R., Salet, C., Balzani, V., *J. Am. Chem. Soc.* (1976) **98**, 3722.
23. Mahon, C., Reynolds, W. L., *Inorg. Chem.* (1967) **6**, 1927.
24. Harrigan, R. W., Crosby, G. A., *J. Chem. Phys.* (1973) **59**, 3468.
25. Hipps, K. W., Crosby, G. A., *J. Am. Chem. Soc.* (1975) **97**, 7042.
26. Bock, C. R., Meyer, T. J., Whitten, D. G., *J. Am. Chem. Soc.* (1975) **97**, 2909.
27. Creutz, C., Sutin, N., *Inorg. Chem.* (1976) **15**, 496.
28. Memming, R., *Photochem. Photobiol.* (1972) **16**, 325.
29. Memming, R., Kursten, G., *Ber. Bunsenges. Phys. Chem.* (1972) **76**, 4.
30. Forster, L. S., in "Concepts of Inorganic Photochemistry," A. W. Adamson, P. D. Fleischauer, Eds., p. 1, Wiley–Interscience, New York, 1975.
31. Hager, G. D., Watts, R. J., Crosby, G. A., *J. Am. Chem. Soc.* (1974) **97**, 7037.
32. See footnote 47 in Lin, C.-T., Böttcher, W., Chou, M., Creutz, C., Sutin, N., *J. Am. Chem. Soc.* (1976) **98**, 6536.
33. Saji, T., Aoyagui, S., *J. Electroanal. Chem.* (1975) **63**, 31.
34. Young, R. C., Keene, F. R., Meyer, T. J., *J. Am. Chem. Soc.* (1977) **99**, 2468.
35. Marcus, R. A., *J. Chem. Phys.* (1965) **43**, 679.
36. Marcus, R. A., *J. Chem. Phys.* (1965) **43**, 2654.
37. Biedermann, G., Silber, H. B., *Acta Chem. Scand.* (1973) **27**, 3761.
38. Latimer, W. M., "Oxidation Potentials," Prentice–Hall, Englewood Cliffs, NJ, 1952.
39. Swift, E. H., "A System of Chemical Analysis," p. 542, Prentice–Hall, New York, 1949.
40. Hoselton, M. A., Lin, C.-T., Schwarz, H., Sutin, N., *J. Am. Chem. Soc.* (1978) **100**, 2383.
41. Toma, H. E., Creutz, C., *Inorg. Chem.* (1977) **16**, 545.
42. Creutz, C., *Inorg. Chem.* (1978) **17**, 1036.
43. Maestri, M., Grätzel, M., *Ber. Bunsenges. Phys. Chem.* (1977) **81**, 504.
44. Creutz, C., Sutin, N., *J. Am. Chem. Soc.* (1977) **99**, 241.
45. Baker, B. R., Orhanovic, M., Sutin, N., *J. Am. Chem. Soc.* (1967) **89**, 722.
46. Meyer, T. J., Taube, H., *Inorg. Chem.* (1968) **7**, 2369.
47. Brown, G., Krentzien, H., Taube, H., cited by H. Taube in "Bioinorganic Chemistry—II," *Adv. Chem. Ser.* (1977) **162**, 127.

48. Gordon, B. M., Williams, L. L., Sutin, N., *J. Am. Chem. Soc.* (1963) **83**, 2061.
49. Ford-Smith, M. H., Sutin, N., *J. Am. Chem. Soc.* (1961) **83**, 1830.
50. Brunschwig, B., unpublished data.
51. Przystas, T. J., Sutin, N., *J. Am. Chem. Soc.* (1973) **95**, 5545.
52. Ballardini, R., Varani, G., Scandola, F., Balzani, V., *J. Am. Chem. Soc.* (1976) **98**, 7432.
53. Chou, M., Creutz, C., Sutin, N., *J. Am. Chem. Soc.* (1977) **99**, 5615.
54. Candlin, J. P., Halpern, J., Trimm, D. L., *J. Am. Chem. Soc.* (1964) **86**, 1019.
55. Faraggi, M., Feder, A., *Inorg. Chem.* (1973) **12**, 236.
56. Carlyle, D. W., Espenson, J. H., *J. Am. Chem. Soc.* (1968) **90**, 2272.
57. Brown, G. M., Clark, W. D. K., unpublished data.
58. Zwickel, A., Taube, H., *J. Am. Chem. Soc.* (1961) **83**, 793.
59. Doyle, J., Sykes, A. G., *J. Chem. Soc., A* (1968) 2836.
60. Nordmeyer, F., Taube, H., *J. Am. Chem. Soc.* (1968) **90**, 1163.
61. Dockal, E. R., Gould, E. S., *J. Am. Chem. Soc.* (1972) **94**, 6673.
62. Endicott, J. F., Taube, H., *J. Am. Chem. Soc.* (1964) **86**, 1686.
63. Faraggi, M., Feder, A., *Inorg. Chem.* (1973) **12**, 236.
64. Gaunder, R. G., Taube, H., *Inorg. Chem.* (1970) **9**, 2627.

RECEIVED September 20, 1977.

ic Analogs of Ru(bipy)$_3^{2+}$

2

Light-Induced Electron Transfer Reactions of Hydrophobic Analogs of Ru(bipy)$_3^{2+}$

PATRICIA J. DELAIVE, J. T. LEE, H. ABRUÑA,
H. W. SPRINTSCHNIK, T. J. MEYER, and DAVID G. WHITTEN[1]

Department of Chemistry, University of North Carolina, Chapel Hill, NC 27514

> *Light-induced electron transfer reactions of several Ru(II) complexes, RuL$_3^{2+}$, (L is* [bipyridine with COOR COOR substituents] *; R = isopropyl, cyclohexyl, benzyl, α-naphthyl, or dihydrocholesteryl) have been investigated. Absorption and luminescence spectra, luminescence lifetimes, and redox properties of these complexes are similar to those of Ru(bipy)$_3^{2+}$. For complexes in which the ruthenium cation is shielded by a hydrophobic barrier, the excited-state electron transfer quenching process and subsequent reactions are modified compared with those of Ru(bipy)$_3^{2+}$. Neutral electron donors or acceptors quench several of the RuL$_3^{2+}$ complex excited states in organic solvents at near diffusion-controlled rates; positively charged acceptors quench some of the RuL$_3^{2+}$ excited states at sharply reduced rates. Upon reductive quenching by triethylamine or N,N-dimethylaniline, RuL$_3^+$ is formed as an isolable but high-energy product.*

The quenching of excited states of transition metal complexes such as tris (2,2′-bipyridine)ruthenium(II)$^{2+}$ (Ru(bipy)$_3^{2+}$) (1) and related compounds by electron transfer reactions is by now a very well established phenomenon (1–7). It has been shown, for example, that Ru(bipy)$_3^{2+*}$ can be quenched efficiently by both electron donors and acceptors. Thus, the presence of a number of closely spaced oxidation states in many transition metal complexes often permits a wider range of excited-state redox processes than can be obtained for organic molecules, where, although electron transfer quenching frequently occurs, the

[1] Senior author.

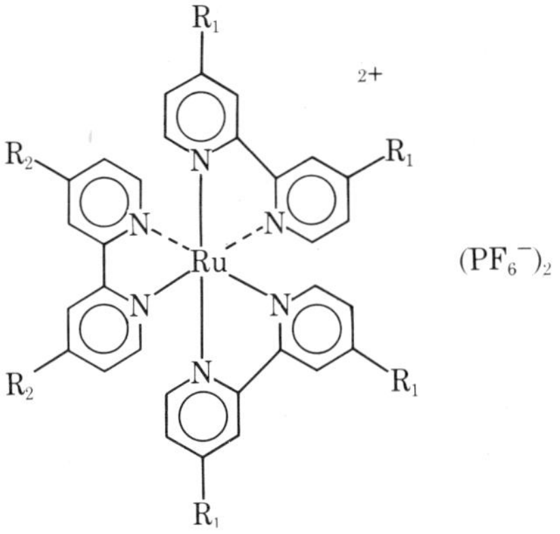

1

much greater separation of oxidation states usually allows an excited organic molecule to serve as either an electron donor or an electron acceptor but not both. In addition, the redox potentials of metal complexes often can be varied systematically by changes in ligand or metal. The presence of redox sites on both metal and ligand offers additional possibilities not available for either simple metal ions or organic molecules. The quenching of an excited state by electron transfer to produce high-energy redox products (Schemes 1 and 2) has been demonstrated to be an efficient energy conversion process in several cases (*1–7*); however, in most instances, any practical utility of these reactions is limited by the subsequent occurrence of very rapid reverse electron-transfer processes which produce the ground states of the starting reagents in an energy-wasting process (*2,6,7*). We have conducted investigations of light-induced electron transfer reactions of metal complexes in solution and in organized media such as micelles and monolayer

Scheme 1. *Oxidative quenching of metal complex excited states*

$$RuL_3^{2+*} + Ox^n \rightarrow [RuL_3^{3+}, Ox^{n-1}] \rightarrow RuL_3^{3+} + Ox^{n-1}$$
$$\downarrow$$
$$RuL_3^{2+} + Ox^n$$

Scheme 2. *Reductive quenching of metal complex excited states*

$$RuL_3^{2+*} + Red^n \rightarrow [RuL_3^{+}, Red^{n-1}] \rightarrow RuL_3^{+} + Red^{n-1}$$
$$\downarrow$$
$$RuL_3^{2+} + Red^n$$

assemblies (8, 9). Our studies in organized media have necessitated the preparation of water-insoluble, surfactant metal complexes; the synthesis of suitable complexes can be obtained by the substitution of a 2,2'-bipyridine ligand by the fatty alcohol esters of 2,2'-bipyridine-4,4'-dicarboxylic acid (9). In the course of these syntheses, we were able to prepare complexes in which a Ru^{2+} ion is surrounded by three hydrophobic bipyridine ligands. These complexes, although containing the charged $Ru(bipy)_3^{2+}$ core, are water insoluble but highly soluble in most organic solvents. We felt that these "hydrophobic" Ru(II) complexes should be attractive candidates for solution studies of light-induced electron transfer reactions because of their enhanced solubility in less polar organic solvent and also because of the possibility that the hydrophobic shield surrounding the polar core might impart a selectivity to electron transfer processes. This paper reports results of our studies of light-induced electron transfer reactions of several hydrophobic Ru(II) complexes with a variety of electron donors and acceptors. The results indicate that in some cases significant retardation of electron transfer reactions can occur and that the coupling of this retardation with rapid further reaction of one of the products of light-induced electron transfer can provide a route to effective bypassing of the energy-wasting reverse electron transfer step and to appreciable energy storage through the resulting redox products.

Experimental Section

Preparation and Purification of Materials. Spectroquality acetonitrile was used without further purification in most cases; when it was necessary to eliminate excess water, the acetonitrile was distilled in a closed system prior to use. Isobutyronitrile was distilled from potassium permanganate and was used within one month of purification. p-Dinitrobenzene (Aldrich) was purified by vacuum sublimation at 110°C. o-Dinitrobenzene was obtained from Aldrich and was recrystallized from ethanol. Commercial N,N-dimethylaniline was vacuum distilled prior to use. Triethylamine was purified by distillation over potassium hydroxide, followed by distillation from sodium. N,N'-Dimethyl-4-4'-bipyridine hexafluorophosphate was prepared by the method of R. Young (10). The substituted bipyridine ligands were prepared by the method reported by Sprintschnik et al. (9). The hydrophobic ruthenium complexes were prepared as reported in the literature (9) except for the isopropyl ester "spider" complex (2), which was synthesized in the following manner. Commercial ruthenium trichloride trihydrate (Alfa Products) was refluxed in isopropyl alcohol with a threefold excess of the isopropyl ester of 4,4'-dicarboxy-2,2'-bipyridine for five days. The reaction was followed spectrophotometrically by observing the appearance of the absorption peak at 470 nm. The solution was added to an excess of water and the unreacted cis-chlorobisbipyridylruthenium complex was precipitated.

The solution was filtered and ammonium hexafluorophosphate was added to the remaining aqueous solution. Complex 2 precipitated out and was filtered. The solid was collected, dissolved in acetone, and filtered, and the acetone was allowed to evaporate. The solid was purified by hplc on a silica gel column and gave the characteristic absorption spectrum. Analysis was done by Integral Microanalytical Laboratories of Raleigh, NC, and gave the following results: C 46.73% (calcd. 47.13%); H 4.69% (calcd. 4.39%); N 5.61% (calcd. 6.11%).

Spectroscopy. UV and visible spectra were recorded on either a Cary 17-I spectrophotometer or a Perkin–Elmer 576-St spectrophotometer. Luminescence excitation and emission spectra were recorded on an Hitachi–Perkin–Elmer MPF-2A spectrofluorimeter equipped with a red sensitive Humamatsu R-446 photomultiplier tube.

Conventional Flash Photolysis. The apparatus used in this investigation is described in detail elsewhere (10). Corning glass filters were placed between the flash tube and the sample cell to insure photolysis in the spectral region desired. The samples were rigorously degassed by several cycles of the freeze–pump–thaw degas method and were sealed under vacuum.

Quantum Yield Determinations. All quantum yields were obtained by irradiating samples in borosilicate glass test tubes in a merry-go-round apparatus. The samples were freeze–pump–thaw degassed, sealed under vacuum, and irradiated with the 4540 Å emission line of a Hanovia medium pressure mercury lamp, using Corning filters 3-73 and 5-58. The incident light intensities were measured by the Reinecke's Salt Actinometer.

Electrochemical Measurements. Spectrograde (MCB) acetonitrile was dried and stored over Davidson 4Å molecular sieves. Reagent grade butyronitrile (Eastman) was distilled from P_2O_5 and stored over activated alumina. Tetra-n-butylammonium hexafluorophosphate, TBAH, was prepared according to Meyer et al. (11). Tetraethylammonium perchlorate (Eastman) was recrystallized from water and dried under vacuum for 72 hr.

The electrochemical experiments were performed using a Princeton Applied Research Model 173 Potentiostat/Galvanostat and a Princeton Applied Research Model 175 Universal Programmer as a signal generator. For cyclic voltammetric experiments, a three-compartment cell was used. The working electrode was a platinum disc (area $\approx$ .13cm^3) and the counter electrode was a platinum grid. For thin-layer spectroelectrochemistry experiments, a cell of design similar to that of Piljac and Murray (12) was used. A Bausch and Lomb Model 210 UV spectrophotometer was used for spectral measurements. All solutions were purged with solvent-saturated nitrogen that had been passed through chromous scrubbers. All potentials are reported vs. the NaCl-saturated calomel electrode. No IR compensation was used in any of the experiments.

Results and Discussion

The hydrophobic Ru(II) complexes **2–9** used in this study have absorption and emission spectra in nitrile solvents very similar to those

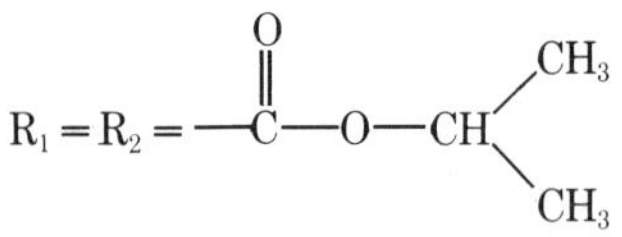

$$R_1 = R_2 = -\overset{\overset{O}{\|}}{C} - O - CH\begin{smallmatrix}CH_3\\CH_3\end{smallmatrix}$$

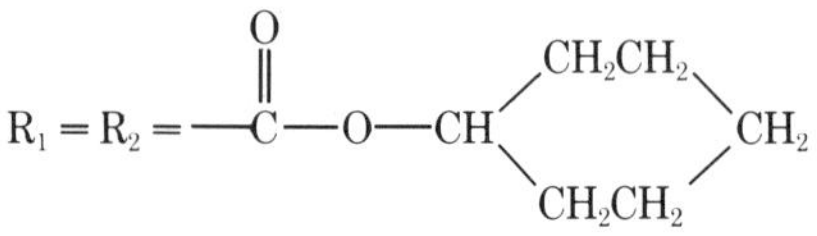

2

$$R_1 = R_2 = -\overset{\overset{O}{\|}}{C} - O - CH\begin{smallmatrix}CH_2CH_2\\ \\ CH_2CH_2\end{smallmatrix}CH_2$$

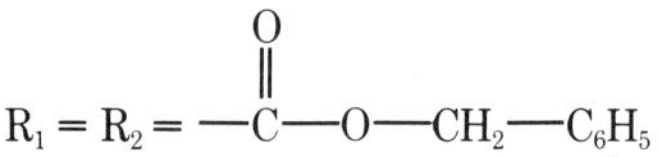

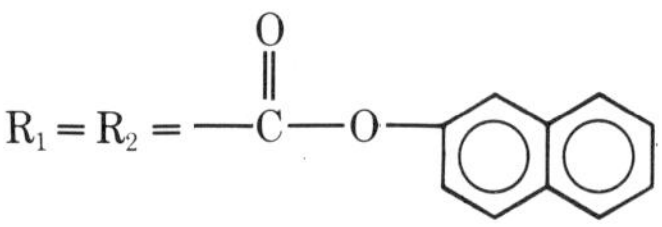

3

$$R_1 = R_2 = -\overset{\overset{O}{\|}}{C} - O - CH_2 - C_6H_5$$

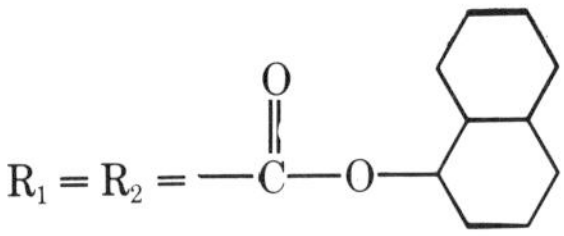

4

$$R_1 = R_2 = -\overset{\overset{O}{\|}}{C} - O - \text{(naphthyl)}$$

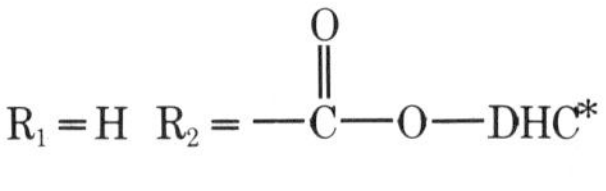

5

$$R_1 = R_2 = -\overset{\overset{O}{\|}}{C} - O - \text{(decalinyl)}$$

6

$$R_1 = H \quad R_2 = -\overset{\overset{O}{\|}}{C} - O - DHC^*$$

7

$$R_1 = -\overset{\overset{O}{\|}}{C} - O - DHC \quad R_2 = H$$

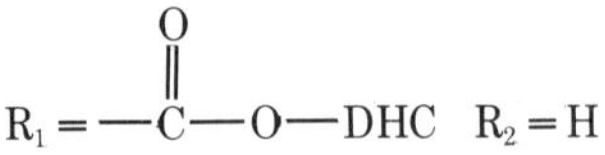

8

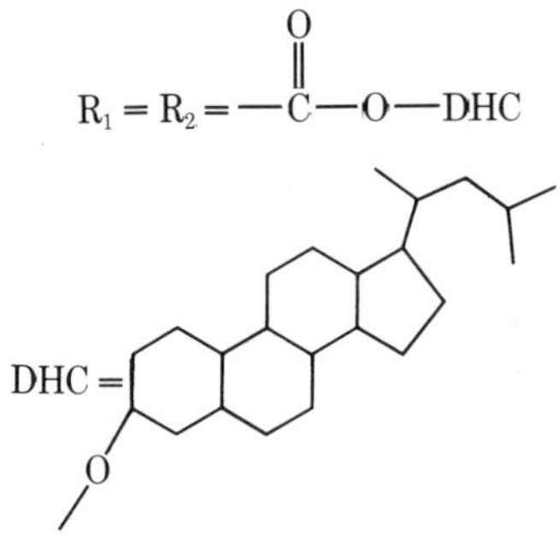

$$R_1 = R_2 = -\overset{\overset{O}{\|}}{C}-O-DHC$$

DHC = [steroid structure]

9

of Ru(bipy)$_3^{2+}$ (**1**). Although there are small red shifts in both the absorption and emission maxima compared with those of **1**, all of the "spider complexes" containing three 4,4'-dicarboxy-2,2'-bipyridine ligands have excited state energies close to 50 kcal/mol, as indicated in Table I. Lifetimes and λmax values for the luminescence of complexes **2–9** also are listed in Table I. The unsymmetrical complexes **7** and **8** have red-shifted luminescence and slightly lower excited state energies. Lifetimes of the hydrophobic complexes are similar to that of **1**, and the intensities of emission for all complexes are comparable at least qualitatively with that of **1**, suggesting comparable quantum efficiencies for luminescence.

Electrochemical oxidations and reductions of several of the complexes were carried out in acetonitrile. As might be expected on the basis of substitution of the bipyridine ligand with the electron-withdrawing carboxyl groups, the potentials for oxidation of the hexacarboxy complexes were shifted by ca. 0.3 V more anodic compared with that of **1** while the corresponding potentials for reduction of the di-cation were shifted

Table I. Luminescence Maxima, Lifetimes, and Excited State Energies for Hydrophobic Ru(II) Complexes

Complex	τ_0 (sec)	$\lambda_{excitation}$ (nm)	$\lambda_{emission}$ (nm)	Excited State Energy (kcal/mol)
2[a]	2.39 × 10^{-6}	466	626	51.3
3[a]	2.21 × 10^{-6}	467	627	50.9
4[a]	1.93 × 10^{-6}	467	634	50.8
5[a]	2.19 × 10^{-6}	468	631	51.3
6[a]	2.21 × 10^{-6}	467	631	51.1
7[b]	2.00 × 10^{-6}	468	666	48.8
8[b]	2.10 × 10^{-6}	468	641	49.9
9[b]	2.14 × 10^{-6}	468	636	50.7

[a] Solvent acetonitrile.
[b] Solvent isobutyronitrile.

Table II. Redox Couples for Hydrophobic Ruthenium Complexes

Complex[a]	$E_{1/2(+3/+2)}$ V vs. SCE	$E_{1/2(+2/+1)}$ V vs. SCE
1	1.29	−1.32
2	1.59	−0.9
3	1.53	−0.91
4	1.55	−0.95
7	1.34	−0.99

[a] Acetonitrile solution with 0.1M tetra-n-butylammonium hexafluorophosphate as supporting electrolyte.

by ca. 0.4 V to more anodic values. For the unsymmetrical complex, **7**, the values for both redox couples are intermediate between those for **1** and the tris-substituted complexes **2, 3,** and **4**. Interestingly, it was found that nitrile solutions of **5, 6, 8,** and **9** were electrochemically inactive under conditions where **1** and the other complexes were easily oxidized and reduced. Presumably, the large hydrophobic shield in these complexes inhibits approach of the electroactive portion of the complex to the platinum electrode surface. Values for the redox potentials for these complexes are listed in Table II.

For **1**, it has been shown that both the oxidizing and reducing properties of the excited state are enhanced by the approximate magnitude of the relaxed excited state energy (2.10 V) (*2, 3, 4*). Figure 1 gives a diagram showing expected formal reduction potentials for the various ground and excited state couples involving **1–9**. In line with the potentials estimated for the 2+*/1+ and 3+/2+* couples, it is found that the luminescence of the hydrophobic complexes **2–9** is quenched by a wide range of electron donors and acceptors.

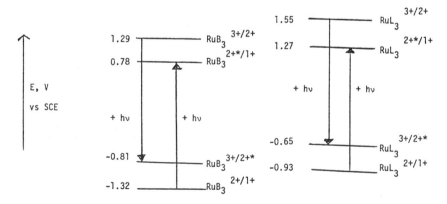

Figure 1. Formal redox relationships for ground and excited states of ruthenium complexes: left side, Ru(bipy)$_3$; right side, hydrophobic complexes

Quenching by Neutral Acceptors. Several neutral acceptors which have been previously shown to quench the luminescence of 1 at near-diffusion controlled rates (2, 3) were also found to quench the luminescence of hydrophobic complexes 2–9. Quenching constants for these three quenchers—tetracyanoethylene, o-dinitrobenzene, and p-dinitrobenzene, for which reduction potentials are 0.24, -0.81, and -0.69 V respectively—are given in Table III. For tetracyanoethylene, the redox state [RuL$_3^{3+}$, Ox$^-$] formed by electron transfer from the ruthenium complex to the oxidant can be calculated from the potentials $E_{1/2}(\text{RuL}_3^{2+}/\text{RuL}_3^{3+})$ − $E_{1/2}$ (Ox$^-$/Ox) to be 24 kcal/mol for 1. For the hexacarboxy complexes 2–6 and 9, values in the range 29.5–31.1 kcal/mol can be calculated from the reduction potential data while for 7, a value of 25.4 kcal/mol is indicated. Clearly for all of these complexes the excited state quenching

Table III. Quenching of Hydrophobic Ruthenium Complex Luminescence by Neutral Electron Acceptors

k_q, $M^{-1}\ sec^{-1\,b}$

Complex[a]	Tetracyanoethylene	o-Dinitrobenzene	p-Dinitrobenzene
1	1.62 × 10^{10}	3.10 × 10^9	6.6 × 10^9
2	n.d.[c]	n.d.	8.1 × 10^7
3	8.8 × 10^9	1.9 × 10^6	6.0 × 10^7
4	4.0 × 10^9	2.3 × 10^6	7.1 × 10^7
5	8.8 × 10^9	2.5 × 10^6	6.0 × 10^7
6	7.1 × 10^9	2.0 × 10^6	5.0 × 10^7
7	3.0 × 10^9	4.8 × 10^6	1.3 × 10^8
8	4.6 × 10^9	2.8 × 10^6	8.3 × 10^7
9	3.1 × 10^9	1.8 × 10^6	5.5 × 10^7

[a] Concentration ~ 10^{-5}M.
[b] Degassed solutions in acetonitrile with 0.1M TEAP at 25°C.
[c] Not determined.

by electron transfer to tetracyanoethylene is a highly favored process, and it would be expected that values near the diffusion-controlled limit should be obtained if access of the quencher to the electroactive site were unhindered. For complexes 3, 4, 5, and 6, where the hydrophobic groups are relatively small, the rates of quenching are about one-half that obtained for 1 with the same quencher. For the dihydrocholesteryl-substituted complexes, there is a somewhat larger decrease in quenching rate constants such that the hexa-substituted complex 9, which would be expected to be highly hindered, is quenched with a rate about one-fifth that for 1. For o-dinitrobenzene and 1, the free energy change for the quenching step is estimated to be nearly equal to the excited-state energy and a lower than diffusion-controlled value for the quenching constant is obtained. Since the hydrophobic complexes are less easily oxidized, the redox state should lie at least 0.25–0.3 V above the excited state such that lower quenching rates are expected even in the absence of shielding by the hydrophobic groups. (The excited state energies listed in Table I are obtained by extrapolating to the onsets of absorption and emission. Although these are certainly reasonable estimates, there is obviously much more uncertainty in these data than in the redox potentials.) From the plot obtained with a variety of neutral electron acceptors (2, 3), it can be estimated that "normal" quenching rate for 2–6 and 9 should be in the range 2×10^6–8×10^6 M^{-1} sec^{-1}. The values actually measured fall in this range, and there appears to be little indication of any trend within the series of compounds studied to date with o-dinitrobenzene. For p-dinitrobenzene and 1, the redox products are estimated to lie 0.12 V below the excited state of 1; for the symmetrical hydrophobic complexes, the redox state should lie 0.13–0.18 V above the complex excited states, and the value for "unhindered" quenching constants would be expected to be in the range 2×10^8–7×10^8 M^{-1} sec^{-1}. ("Anticipated" values for quenching constants were obtained from the plot in Ref. 16 using the excited state energies and redox potentials for the hydrophobic complexes.) The values obtained are slightly lower (one-half to one-fifth the "expected" value), and once again, no clear cut trend can be established for the different shielding groups. As was observed for quenching of excited 1 by the same quenchers, there are no transients detectable by microsecond flash photolysis occurring as a consequence of the quenching and no permanent photobleaching. Apparently, electron transfer quenching in these cases produces a very short-lived ion pair that rapidly decays back to starting materials.

Quenching by Positively Charged Acceptors. Previously we have shown that several quaternary salts of N-heteroaromatics such as the dimethylbipyridinium ions, 1,2-bispyridylethylene salts, and N,N-dimethylphenanthrolines quench the luminescence of 1 with concurrent production of the monocations and Ru(bipy)$_3^{3+}$, as shown in Scheme 1. As indicated earlier, the products rapidly undergo back electron transfer, a process readily observable by flash photolysis. The di-cation Paraquat^{2+} (N,N'-dimethyl-4,4'-bipyridine) was used as a quencher for the luminescence of complexes 2–9. In each case, qualitatively similar behavior was observed, but the quenching rates are substituent dependent and substantially slower than those obtained for 1. Table IV gives quenching constants and some rate constants obtained by flash photolysis for the back electron transfer process.

For Paraquat^{2+}, the quenching ate constants for 2, 3, and 5 are lower by about one order of magnitude than that for 1. The reduction in rates could be attributed in part to less favorable energetics for the electron transfer process for the hydrophobic complexes since the oxidation is, as mentioned previously, less favorable by ca. .25–.3 V. However, for all three complexes, the electron transfer process is estimated to require 46–47 kcal/mol which is 4–5 kcal/mol below the excited state energy. Thus, it appears rather surprising that the measured rate constants are nearly two orders of magnitude below the diffusion-controlled value. Examination of previously obtained values for a series of similar di-cations and 1 (13) indicates that the "curve" for log kq vs. quencher reduction potential is "displaced" compared with the curve obtained for the neutral quenchers with consistently lower-than-expected values for the di-cations. Nonetheless, the values for 2, 3, and 5 with Paraquat^{2+} are below the value estimated for quenching of 1 under "isoenergetic" conditions (13). The value expected for $\Delta G = \Delta E_{\text{redox products}} - E_T = -4$ to -5 kcal/mol would be ca. 6×10^8 M^{-1} sec^{-1}, and the actual values

Table IV. Quenching of Hydrophobic Ruthenium Complex Luminescence by Positively Charged Electron Acceptors

Complex[a]	k_q, M^{-1} sec^{-1}[b] Paraquat^{2+}	$k_B PQ^{2+}$[c]
1	2.8×10^9	8.1×10^9
2	1.2×10^8	1.8×10^9
3	1.7×10^8	3.5×10^9
4	n.d.[d]	
5	1.6×10^8	
6	n.d.[d]	
7	3.5×10^7	
8	3.6×10^7	
9	1.7×10^7	1.3×10^9

[a] Concentration ~ $10^{-5}M$.
[b] Degassed solutions in 1:1 acetonitrile:isobutyronitrile with 0.1M TEAP at 25°C.
[c] Degassed solutions in 1:1 acetonitrile:isobutyronitrile, photolyzed at 605 nm.
[d] Not determined.

are only one-third to one-sixth of this. The quenching values for the three dihydrocholesteryl-substituted complexes are even lower than those for 2, 3, and 5. For 7, which has a lower excited state energy but a more favorable redox potential for the $+3/+2$ couple, ΔG is estimated to be ca. -7.5 kcal/mol; the value measured for kq is between one and two orders of magnitude below the "anticipated" value. Complex 9 should have similar excited state energy and reduction potentials of 2, 3, and 5; however, the rate constant for quenching of 9 by Paraquat^{2+} is one order of magnitude lower than for 3 or 5 and approximately 1/40th of that expected in the absence of shielding by the hydrophobic groups. The values obtained indicate that shielding by the hydrophobic groups affects kq for all of the complexes in the case of quenching by Paraquat^{2+} and reveal that the effect is both more pronounced and more substituent-dependent for the positively charged quencher than for neutral acceptors such as nitroaromatics or tetracyanoethylene.

Some approximate rate constants for the Paraquat^{1+} + complex^{3+} back reaction are given in Table IV. The values for complexes 2, 3, and 9 are lower than that for 7, but the effect appears much less than that for the forward quenching process. Nonetheless, the effect is noteworthy since here the reaction is energetically highly favored in each case. (ΔG for the back reaction is 0.2–0.3 V more favorable for the hydrophobic complexes compared with 1. A possible complication worth noting is that these reactions are all in the "nonadiabatic" free-energy region where theory (14, 15, 16, 17) predicts a decrease which until now has not been observed in similar reactions.)

Quenching and Reaction with Neutral Electron Donors. It has previously been shown that various electron donors can reduce excited

Table V. Quenching of Ru(II) Complex Excited States by Amines

Complex[a]	k_q, M^{-1} sec^{-1}[b]	
	N,N-Dimethylaniline	Triphenylamine
1	7.1 × 10^7	
2	3.7 × 10^9	
3	4.5 × 10^9	1.5 × 10^9
4	4.7 × 10^9	
5	4.1 × 10^9	
6	4.3 × 10^9	
7	3.7 × 10^8	
8	2.1 × 10^9	
9	2.3 × 10^9	3.5 × 10^9

[a] Concentration ~ 10^{-5}M.
[b] Degassed solutions in acetonitrile with 0.1M TEAP at 25°C.

states of 1 and related complexes by electron transfer (7, 10, 18). Since complexes 2–9 should be even better electron acceptors than 1, it appeared of interest to investigate quenching by electron donors. A particularly attractive potential aspect of quenching by neutral electron donors was suggested by the results presented above with electron acceptors. Since the retardation of quenching was found to be considerably greater with positively charged acceptors than with one neutral quencher, it appeared possible that the quenching process with neutral donors might also be relatively rapid but that a pronounced reduction in the corresponding back electron-transfer process (Scheme 2) might occur. Toward this end, we have examined quenching of the hydrophobic complexes by a series of amines. Table V gives some values for quenching constants obtained with N,N-dimethylaniline (DMA) and triphenylamine (TPA). For DMA the quenching constant obtained with 1 (10, 18) is much lower than those obtained for the hydrophobic complexes. The more rapid rates for the hydrophobic complexes are consistent with the more favorable free energy change for the electron-transfer quenching process; the reaction with the hydrophobic complexes is more favorable by ca. 9 kcal/mol, and ΔG for the excited state process becomes negative for these substrates. The relative rate differences in the series are small but more or less parallel to those obtained for quenching of the same complexes by electron acceptors. The isopropyl complex 2 is quenched a little slower than are the cyclohexyl (3), benzyl (4), or naphthyl (5) complexes, while the dihydrocholesteryl complex (9) is quenched even more slowly. Not enough data are available for quenching by triphenylamine to make comparisons. Formation of transient products having the spectra expected for the reaction:

$$Ru^{2+*} + TPA \rightarrow Ru^+ + TPA^+$$

was observed with both 3 and 9. The products returned to the starting materials with rate constants of 1.5×10^9 and 3.5×10^9 M^{-1} sec^{-1} for 3 and 9, respectively.

Back reaction of DMA$^+$ with Ru(bipy)$_3^{3+}$ has been observed to occur in acetonitrile with a rate constant of 4.1×10^9 M^{-1} sec^{-1} (10, 18). In contrast, attempts to observe the back reactions of DMA$^+$ with several of the reduced hydrophobic complexes in acetonitrile following excited state quenching were unsuccessful since only a permanent bleaching could be observed with these solutions following flash or steady-state illumination. (However, we find back reaction does occur with several p-substituted DMA derivatives.) The photochemical changes that occur on irradiation of mixtures of DMA and the hydrophobic complexes appear complicated, and no simple conclusions regarding the products formed can be made. However, light-induced reactions of the complexes with triethylamine, which would be expected to have at least similar initial steps, have been found to be relatively straight forward. Although quenching of the entire series of complexes with triethylamine has not been investigated, the luminescence of 2 is quenched with a rate constant of 1.4×10^8 M^{-1} sec^{-1}, which appears reasonable from a consideration of redox potentials. In contrast, quenching of the luminescence of 1 can be observed only at very high triethylamine concentrations since the electron-transfer process is obviously energetically unfavorable ($E_{p,a}$ [Et$_3$N → Et$_3$N$^+$ +e] = 1.04 V (19), electrochemically irreversible). Quenching of the luminescence of 1 by triethylamine produces spectral changes consistent (10, 18, 20) with production of Ru(bipy)$_3^+$ and N(Et$_3$)$^+$; the products rapidly undergo back electron transfer so that no permanent changes occur. Even prolonged irradiation of 1 in the presence of triethylamine leads to no bleaching or spectrally detectable product formation. In contrast, quenching of the luminescence of **2, 3,** and **9** results in no detectable back reaction. In dry degassed acetonitrile solutions, permanent changes occur on irradiation which are consistent with an irreversible reduction of the ruthenium complex. For example, irradiation of $5 \times 10^{-4}M$ of **2** in dry acetonitrile containing $0.3M$ triethylamine leads to spectral changes shown in Figure 2. The product spectrum is nearly identical to that of the product of one-electron reduction of **2** generated by electrochemical reduction. The product is stable over a period of days in dry acetonitrile but rapidly reverts to the starting spectrum when air or water are introduced. The contrast in behavior of **1** and **2** on irradiation in the presence of triethylamine is rather striking. With both Ru(II) complexes, reductive quenching of the excited state by the amine leads to the reduced Ru(II)$^{+1}$ species. In the case of **1** rapid back reaction occurs while with **2** the back reaction must be slowed sufficiently to allow reaction of the triethylamine radical cation to provide an alternate path such that the high energy-reduced ruthenium complex survives. Two paths for rapid reaction of the triethylamine radical cation that can be proposed are outlined

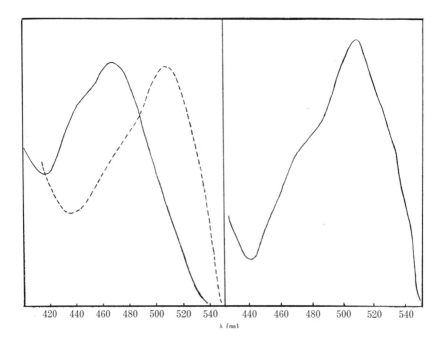

*Figure 2. Left trace: spectrum of **2** before (solid line) and after (dashed line) photolysis in triethylamine:acetonitrile. Right trace: spectrum of the one-electron reduction product of **2** in acetonitrile.*

$$\overset{+}{\text{N}}\text{Et}_3 + \text{CH}_3\text{CN} \rightarrow \text{H}\overset{+}{\text{N}}\text{Et}_3 + \cdot\text{CH}_2\text{CN} \qquad (1)$$

$$2 \cdot \text{CH}_2\text{CN} \rightarrow \begin{array}{c} \text{CH}_2 - \text{CN} \\ \text{CH}_2 - \text{CN} \end{array} \qquad (2)$$

$$\cdot\overset{+}{\text{N}}\text{Et}_3 + :\text{NEt}_3 \rightarrow \text{Et}_2\overset{..}{\text{N}}-\overset{.}{\text{C}}\text{HCH}_3 + \text{H}\overset{+}{\text{N}}\text{Et}_3 \qquad (3)$$

$$\text{Et}_2\overset{..}{\text{N}} - \overset{.}{\text{CH}} - \text{CH}_3 \xrightarrow{-e^-} \text{Et}_2\text{N}^+ = \text{CH} - \text{CH}_3 \qquad (4)$$

in Reactions 1, 2, 3, and 4. Both sequences have been proposed in related cases (21, 22). A sequence involving Reactions 3 and 4 has been proposed in the photoreduction of benzophenone and other ketones by amines (23, 24); in this case, the occurrence of Reaction 4 results in reduction of a second ketone, producing a limited quantum yield of 2 for the disappearance of ketone. If Reaction 3 occurs with the ruthenium complex, it would be reasonable to expect that the radical formed could react with a second molecule of complex giving $[\text{Ru(II)L}_3^-]^+$, so that here also a limiting value of $\phi = 2$ might be expected. In either case, thermochemical calculations indicate that formation of the Ru(II)^{+1} results in storage of a substantial fraction of the excitation energy.

Irradiation of **2** in moist acetonitrile containing triethylamine results in no net reduction of the ruthenium complex (vide infra). Sustained irradiation of 470 nm where **2** absorbs leads to the production of acetaldehyde as the major organic product detectable by VPC; presumably it arises through hydrolysis of the imine formed in Reaction 4 (23). A trace of product having a retention time identical to that of succinonitrile is also detectable; however, it appears that reaction via Reactions 3 and 4 is the predominant path for decay of the triethylamine radical cation.

The quantum efficiency for production of reduced **2** on short-term irradiation of $5 \times 10^{-4}M$ solutions of **2** in dry acetonitrile is 0.35 for 470-nm light. Formation of reduced **2** also occurs on irradiation in isobutyronitrile and tetrahydrofuran, with quantum yields of 0.2 and 0.05, respectively. As mentioned previously, similar behavior on irradiation has been observed for **2, 3,** and **9**; however, the reaction appears not to be as spectroscopically well-behaved for the latter two complexes in nitrile solvents, and regeneration of the starting complex on admission of air or water following the photoreduction is incomplete, at least in the dilute solutions studied to date.

The demonstration that the high-energy Ru(II)$^{+1}$ species can be formed in good yield, as shown spectroscopically for **2**, is clearly a promising result. The net reaction involves the use of visible (470 nm) light to effect relatively efficient conversion and storage of energy. The formulation of a practical system for any kind of application would, of course, depend on development of further use of the reduced ruthenium species and a tuning of the reaction to obtain better efficiencies or use of more easily obtainable reagents. As mentioned earlier, the Ru(II)$^{+1}$ species generated from photolysis of **2** is stable in degassed solution but reactive towards a wide variety of reagents including oxygen and water. Although the luminescence of **2** is unquenched by water in acetonitrile, irradiation of **2** in the presence of triethylamine and water leads to no net reduction of the complex. Although an extensive investigation of the permanent products formed on irradiation under these conditions has not yet been completed, it appears that molecular hydrogen is not a significant product in the relatively dilute solutions studied.

The present results with a limited number of shielding groups surrounding the reactive metal complex core suggest that a tuning of redox rates could be possible for a wide variety of systems with possibly even more dramatic results. Although considerably more investigations will be necessary to better elaborate the role of the shielding groups even in this one system, some reasonable inferences can be made with the present data. The observation that even relatively large molecules such as triphenylamine can quench the hydrophobic complexes suggests that there must be either gaps in the shield or enough flexibility to permit quenchers sufficient access to allow electron transfer to occur. The fact that the isopropyl groups are as effective (or even more effective in certain cases) as benzyl, cyclohexyl, or naphthyl groups in reducing quenching rates indicates that the larger groups do not serve as effective "spacers" in holding quenchers away from the complex center. The effectiveness of the isopropyl group may, in fact, indicate that substitution of less rigid groups at or close to the alcohol group produces a more effective blocking of the gaps into which quencher molecules must penetrate. On the other hand, the evidence that quenching by Paraquat^{2+} and the back reaction between NEt$_3^+$ and reduced complex are most affected by the hydrophobic shielding groups of the reactions studied suggests that the shield may be acting at least in part as a dielectric or effective low-dielectric constant solvent in excluding molecules with a high ratio of charge to mass. This suggest that both quenching and back reaction with small hydrophilic reagents such as metal ions may be strongly affected with these complexes. Experiments to further investigate these effects, using both a variety of new substrates and new quenchers, are currently underway.

Acknowledgment

We thank R. C. Young for measuring excited-state lifetimes and D. Salmon for measuring some of the redox couples reported. We are grateful to the National Institutes of Health (Grant GM.15,238) and the National Science Foundation (Grants CHE 76-01074 and CHE-14405) for support of this work.

Literature Cited

1. Balzani, V., Moggi, L., Manfren, M. F., Bolleta, F., Gleria, M., *Science* (1975) **189**, 852.
2. Bock, C. R., Meyer, T. J., Whitten, D. G., *J. Am. Chem. Soc.* (1974) **96**, 4710.
3. Ibid. (1975) **97**, 2909.
4. Creutz, C., Sutin, N., *Inorg. Chem.* (1976) **15**, 496.
5. Creutz, C., Sutin, N., *Proc. Nat. Acad. Sci. USA* (1975) **72**, 2858.
6. Young, R. C., Meyer, T. J., Whitten, D. G., *J. Am. Chem. Soc.* (1976) **98**, 286.
7. Lin, C. T., Sutin, N., *J. Phys. Chem.* (1976) **80**, 97.
8. Whitten, D. G., Hopf, F. R., Quina, F. H., Sprintschnik, G., Sprintschnik, H. W., *Pure Appl. Chem.* (1977) **49**, 379.
9. Sprintschnik, G. H. W., Kirsch, P. P., Whitten, D. G., *J. Am .Chem. Soc.* (1977) **99**, 4947.
10. Young, R. C., Ph.D. dissertation, University of North Carolina, 1977.
11. Callahan, R. W., Brown, G. M., Meyer, T. J., *Inorg. Chem.* (1975) **14**, 1443.
12. Piljac, I., Murray, R. W., *J. Electrochem. Soc.* (1971) **118**, 1758.
13. Gutierrez, A. R., Ph.D. dissertation, University of North Carolina, 1975.
14. Marcus, R. A., *J. Chem. Phys.* (1956) **24**, 966.
15. Ibid. (1965) **43**, 679.
16. Marcus, R. A., Sutin, N., *Inorg. Chem.* (1975) **14**, 213.
17. Hush, N. S., *Prog. Inorg. Chem.* (1967) **8**, 39.
18. Young, R. C., Anderson, C. P., Salmon, D. J., Meyer, T. J., *J. Am. Chem. Soc.* (1977) **99**, 1980.
19. Mann, C. K., Barnes, K. K., "Electrochemical Reactions in Nonaqueous Systems," p. 279, Marcel Dekker, New York, 1970.
20. Creutz, C., Sutin, N., *J. Am. Chem. Soc.* (1976) **98**, 6384.
21. Russell, C. P., *Anal. Chem.* (1963) **35**, 1291.
22. Smith, P. J., Manny, C. K., *J. Org. Chem.* (1969) **34**, 1821.
23. Cohen, S. G., Parola, A., Parsons, G. H., Jr., *Chem. Rev.* (1973) **73**, 141, and references therein.
24. Cohen, S. G., Baumgarten, R. J., *J. Am. Chem. Soc.* (1965) **87**, 2996.

RECEIVED September 20, 1977.

3

Photochemistry of Metal–Isocyanide Complexes and Its Possible Relevance to Solar Energy Conversion

HARRY B. GRAY, KENT R. MANN, NATHAN S. LEWIS, JOHN A. THICH, and ROBERT M. RICHMAN

Arthur Amos Noyes Laboratory of Chemical Physics, California Institute of Technology, Pasadena, CA 91125

> *Formation of $M(CNPh)_5py$ and $M(CNIph)_5py$ occurs upon 436-nm irradiation of $M(CNPh)_6$ and $M(CNIph)_6$ (M = Cr, Mo, W; Iph = 2,6-diisopropylphenyl) in pyridine, with quantum yields decreasing according to $[Cr(CNPh)_6]$ (0.23) ~ $[Cr(CNIph)_6]$ (0.23) > $[Mo(CNPh)_6]$ (0.055) > $[Mo(CNIph)_6]$ (0.022) > $[W(CNPh)_6]$ (0.011) >> $[W(CNIph)_6]$ (0.0003). Irradiation of $M(CNIph)_6$ (M = Cr, Mo, W) at 436 nm in $CHCl_3$ yields the one-electron oxidation products $[M(CNIph)_6]Cl$. Similar irradiation of $M(CNPh)_6$ (M = Mo, W) in $CHCl_3$ gives two-electron oxidation products, $[M(CNPh)_6Cl]Cl$. Irradiation (546 nm) of $[Rh_2(bridge)_4](BF_4)_2$ (bridge = 1,3-diisocyanopropane) in 12M HCl yields $[Rh_2(bridge)_4Cl_2]^{2+}$ and H_2. The quantum yield decreases with decreasing $[H^+]$ in HCl from 0.0079 in 12.8M HCl to 0.00010 in 8.1M HCl; in 9M HBr, Φ = 0.044.*

For several years we have been investigating the spectroscopic properties and the photochemistry of metal complexes containing isocyanide ligands. Our interest in this area can be traced to an early realization that isocyanide complexes of low-valent metals exhibit low-lying metal-to-ligand charge transfer (MLCT) bands and that in many cases these bands fall well below the predicted positions of the lowest d-d transitions. Thus, isocyanide complexes offer an opportunity to study in detail the reactivity of MLCT excited states in situations in which the d-d excited states are considerably more energetic (and, presumably, do not intercede to complicate the reactivity pattern).

When we started our work on the photochemistry of metal isocyanide complexes, we had two types of excited state reactivity in mind. First, we were interested in whether an MLCT state would be prone to undergo substitution and, if so, by what type of mechanism. From the fact that the metal center possesses decreased electron density and the M–L bond generally is not appreciably weakened (in contrast to the situation for $d\pi \rightarrow d\sigma^*$ excited states), an MLCT state might favor an associative substitution pathway:

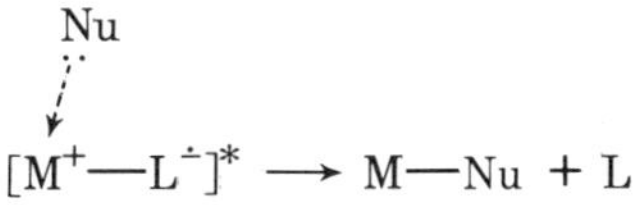

$$[M^+\!-\!L^{\dot-}]^* \xrightarrow{Nu} M\!-\!Nu + L$$

The attacking nucleophilic reagent (Nu:) would have to be positioned near the complex at the time of excitation, as a solvent molecule would be for reactions in solution.

We also expected to observe electron transfer from MLCT excited states to acceptor substrates:

$$[M^+\!-\!L^{\dot-}]^* \, S \longrightarrow M^+\!-\!L + \dot{S}$$

In this area we have been interested especially in designing systems whose MLCT excited states would undergo two electron transfers, as needed, for example, in the reduction of two protons to molecular hydrogen. A very simple idea we had during the course of our work is that the interaction of an electron donor with the MLCT excited species might promote a second electron transfer, according to the following scheme:

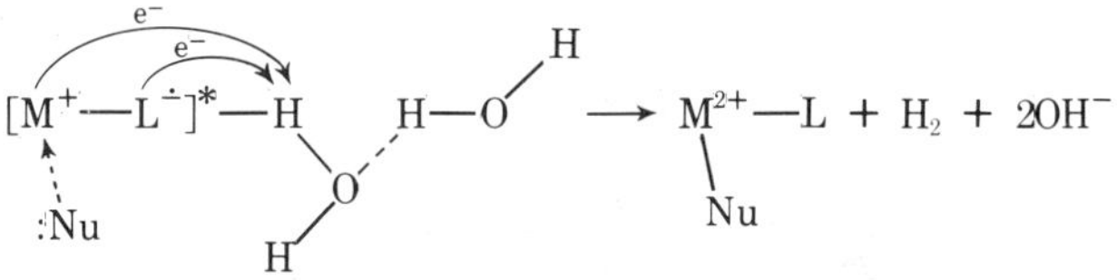

$$[M^+\!-\!L^{\dot-}]^*\!-\!H \quad H\!-\!O\!\diagdown_{H}^{H} \longrightarrow M^{2+}\!-\!L + H_2 + 2OH^-$$
$$:Nu \qquad\qquad\qquad\qquad Nu$$

With these simple ideas providing a background, we now shall proceed to describe what types of photoreactions have, in fact, been observed.

$M(CNPh)_6$ and $M(CNIph)_6$ Complexes

The complexes we have examined in detail are as follows:

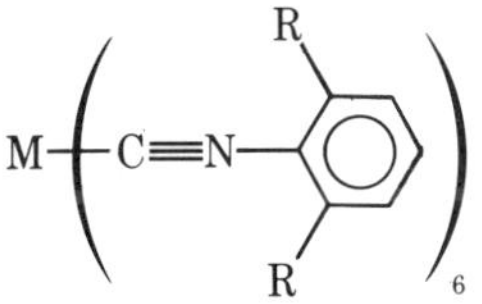

$$M\!\!\left(\!\!-C\!\equiv\!N\!-\!\!\underset{R}{\overset{R}{\bigcirc}}\!\!\right)_6$$

R = H, $M(CNPh)_6$
R = i-Pr, $M(CNIph)_6$

M = Cr, Mo, W

The Cr(0) complexes were prepared by adding excess ligand to chromous acetate, as described by Malatesta (1), and the Mo(0) and W(0) complexes were prepared as follows (2, 3):

$$Mo_2(acetate)_4 \xrightarrow[\Delta]{xs\ L} MoL_6 + \text{other products}$$

$$W(C_2Ph_2)_3CO \xrightarrow[\Delta]{xs\ L} WL_6 + 3C_2Ph_2 + CO$$

We have completed an x-ray crystal structure analysis of $Cr(CNPh)_6$ (3). The CrC_6 core is closely octahedral, with Cr–C and C–N bond lengths of 1.93 and 1.17 Å, respectively (Figure 1). The absorption and emission spectra of the d^6 low-spin $M(CNPh)_6$ and $M(CNIph)_6$ complexes have been studied in solution (4, 5). The results are in Table I. Each complex exhibits an intense absorption system in the 400–500 nm region; in addition, the spectra of the W(0) complexes display a weaker absorption between 500 and 600 nm. The intense absorption system, in each case, is assigned to a spin-allowed $d\pi \to \pi_v^*$ transition (π_v^* is the lowest unoccupied, out-of-plane π^* (CNR) orbital). The MLCT char-

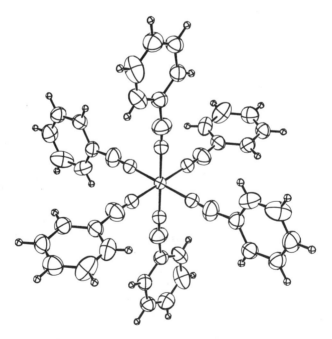

Figure 1. View of $Cr(CNPh)_6$ down the threefold axis

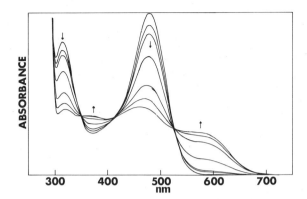

Figure 2. Spectral changes upon visible (fluorescent light) irradiation of a degassed pyridine solution of Cr(CNIph)$_6$ at 25°C. Quantum yields were obtained for irradiations at 436 nm.

acter of this transition was established by comparing band positions in the spectra of analogous Cr(CNR)$_6$ and Mn(CNR)$_6^+$ complexes; the lowest intense system is blue-shifted substantially in the Mn(I) case, as predicted. The weaker 500–600 nm system in the spectra of the W(0) complexes may be attributed to a transition to a spin-orbit component of a $(d\pi)^5(\pi_v{}^*)^1$ triplet state.

All of the ML$_6$ complexes, except Cr(CNPh)$_6$, emit at room temperature in fluid solution; at 77°K in 2-MeTHF glass, Cr(CNPh)$_6$ also emits. We have shown that the lifetime of the Cr(CNIph)$_6$ emission in 2-methylpentane at 77°K is much shorter than that for either Mo(CNIph)$_6$ or W(CNIph)$_6$ (see Table I). We suspect that the emission in the Cr(0) case is a spin-allowed (singlet–singlet) $\pi_v{}^* \to d\pi$ process, whereas the Mo(0) and W(0) emissions originate in a spin-orbit component of the lowest $(d\pi)^5(\pi_v{}^*)^1$ triplet state. Further study of the nature of these emitting states clearly is needed.

Irradiation (436 nm) of the ML$_6$ complexes in pyridine solution results (5) in photosubstitution (Figure 2):

$$\mathrm{ML_6} \xrightarrow[\mathrm{py}]{436 \text{ nm}} \mathrm{ML_5py} + \mathrm{L} \qquad (1)$$

Quantum yield data show interesting trends, as set out below:

M	Φ(L = CNPh)	Φ(L = CNIph)
Cr	0.23	0.23
Mo	0.06	0.02
W	0.01	≤ 0.0003

Table I. Absorption and Emission Spectral Data for ML_6 Complexes

ML_6	Absorption[a]		Emission	
	λ_{max}, nm	$\epsilon \times 10^{-3}$	λ_{max}, nm[b]	τ^c
$Cr(CNPh)_6$	476	40.1	590 (400)	d
	397	65.6		
	321	35.5		
	330	65.6		
$Cr(CNIph)_6$	475	76.0	583 (380)	< 10 nsec
	402	26.1		
	312	64.1		
	242	47.6		
$Mo(CNPh)_6$	475	35.0	559 (450)	d
	382	76.2		
	311	46.0		
	250	55.8		
	234	57.6		
$Mo(CNIph)_6$	475	71.6	568 (450)	40.2 μsec
	435	51.3		
	384	34.1		
	310	79.0		
	245	48.6		
$W(CNPh)_6$[e]	466	31.5	571 (450)	d
	368	81.0		
	315	55.6		
	250	54.8		
	234	55.6		
$W(CNIph)_6$	521	14.5	578 (400)	7.6 μsec
	463	60.0		
	434	44.3		
	373	37.0		
	310	85.7		
	283	44.3		
	250	58.0		

[a] At 300°K in THF solution.
[b] At 77°K in 2-MeTHF glass; excitation wavelength in parentheses.
[c] At 77°K in 2-methylpentane.
[d] Not measured.
[e] Enhanced absorption was observed in solution on the low energy tail of the 466 nm band; at 77°K this resolves into a shoulder at 540 nm.

The Cr(0) photoreactions have a reasonably high quantum yield; this yield is the same for both CNPh and CNIph complexes. The quantum yield data indicate that the Mo(0) and W(0) photosubstitutions are quite different. Not only are the quantum yields much lower but the $M(CNIph)_6$ complexes, that are extremely hindered sterically by the presence of i-Pr groups in the 2,6-phenyl ring positions, have the lowest quantum yields of all. The extreme case of low reactivity is represented by W(CNIph)$_6$, whose quantum yield for photosubstitution is barely measurable.

The results suggest that the excited state mechanism is mainly dissociative in the Cr(0) complexes but that the pathway takes on substantial associative character in the Mo(0) and the W(0) complexes. The observation that $\Phi(W(CNPh)_6) \gg \Phi(W(CNIph)_6)$ is compelling evidence for an associative excited-state substitution mechanism. Inspection of space-filling models shows that [W(CNPh)$_6$]* (assuming, of course, that its excited state structure is not radically different from that of the ground state) is accessible to attack by an incoming nucleophilic reagent, whereas [W(CNIph)$_6$]* most certainly is not. A schematic representation of the proposed associative excited-state substitution involving [W(CNPh)$_6$]* and pyridine is shown in Figure 3.

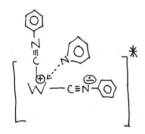

Figure 3. Proposed associative excited state mechanism for the substitution of pyridine for CNPh in F(CNPh)$_6$. Pyridine attack is blocked sterically by the bulky i-Pr groups in W(CNIph)$_6$, providing an explanation for the observation that $\Phi(W(CNPh)_6 \gg \Phi(W(CNIph)_6)$.

The exact nature of the excited states undergoing photosubstitution is still an open question. The reactive Cr(0) states may be MLCT singlets that are energetic enough to lead to Cr–CNR bond cleavage; it also is possible that the reactive Cr(0) states have at least partial $d\pi \rightarrow d\sigma^*$ character, as the ligand field splitting would be expected to be lower in the Cr(0) complexes than in the Mo(0) or W(0) systems. The reactive Mo(0) and W(0) states may be the longer-lived MLCT spin-orbit triplet components, with little or no $d\pi \rightarrow d\sigma^*$ character. As the Mo–C and W–C bond strengths are expected to be greater, the only option open for substitution appears to be an associative one that involves some bond making with the entering group.

Two types of photoredox reactions are observed (5) upon 436-nm irradiation of ML$_6$ complexes in degassed CHCl$_3$ (Figures 4 and 5).

$$M(CNIph)_6 \xrightarrow[CHCl_3]{436 \text{ nm}} M(CNIph)_6^+ \qquad (2)$$

$$\Phi(Cr) \cong \Phi(Mo) \cong \Phi(W) = 0.19(1)$$

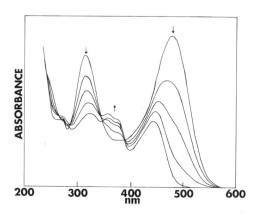

Figure 4. Spectral changes upon visible (fluorescent light) irradiation of a degassed $CHCl_3$ solution of $Cr(CNIph)_6$ at 25°C. Quantum yields were obtained for irradiations at 436 nm.

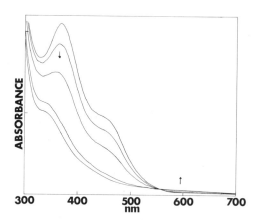

Figure 5. Spectral changes upon visible (fluorescent light) irradiation of a degassed $CHCl_3$ solution of $W(CNPh)_6$ at 25°C. Quantum yields were obtained for irradiations at 436 nm.

Figure 6. Proposed mechanism for the reaction of $[W(CNPh)_6]^*$ with $CHCl_3$. A chlorine atom of $CHCl_3$ binds to the W^+ center in the MLCT excited state, leading to W–Cl bond formation and a net two-electron oxidation to yield $W(CNPh)_6Cl^+$. Such direct chlorine atom attack is not possible in the hindered $[M(CNIph)_6]^*$ complexes, and the result is one-electron outer sphere transfer to form $M(CNIph)_6^+$.

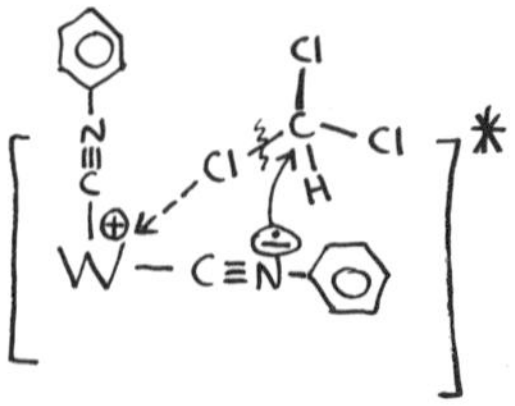

$$M(CNPh)_6 \xrightarrow[CHCl_3]{436 \text{ nm}} M(CNPh)_6Cl^+ \qquad (3)$$

$$\Phi(W) = 0.28(2) > \Phi(Mo) = 0.11(1)$$

The products $Cr(CNIph)_6^+$ and $Mo(CNPh)_6Cl^+$ were isolated and were characterized fully as PF_6^- and Cl^- salts, respectively (3,5). Other $M(CNIph)_6^+$ and $M(CNPh)_6Cl^+$ products were characterized by IR and electronic spectral measurements. The photooxidation of $M(CNPh)_6$ to $M(CNPh)_6Cl^+$ is of special interest, as it provides a concrete example of a net two-electron transfer from $[M(CNPh)_6]^*$ to $CHCl_3$ ($M°$ becomes $M^{II}Cl^+$). We believe that the excited state mechanism is related to the mechanism for pyridine substitution (see Figure 3) in that, shortly after one electron is transferred from $[W(CNPh)_6]^*$ to $CHCl_3$, a chlorine atom may bond directly to W^+, yielding the observed $W(CNPh)_6Cl^+$ product. The excited-state inner electron transfer mechanism that we postulate for this two-electron oxidation is shown in Figure 6.

The $[M(CNIph)_6]^*$ complexes are too hindered to bond a chlorine atom from a $CHCl_3$ (or $CH\overline{Cl}_3$) molecule, so the photoreaction terminates with an outer sphere one-electron transfer, yielding $M(CNIph)_6^+$ products. The quantum yield data are consistent with these mechanistic ideas. In the one-electron process, the excited-state reaction does not involve either breaking a M–L or making a M–Cl bond; the quantum yield probably is limited by the diffusion (and/or breakdown) of $CH\overline{Cl}_3$ from ML_6^+, and this process should not depend significantly on the nature of the central metal. In the $[M(CNPh)_6]^*$ reactions, however, M–Cl bond making will clearly be important in determining how many excited species form products; the expectation that W–Cl bonds are stronger than Mo–Cl bonds accords with the observed quantum yield data.

The observation that a two-electron inner sphere oxidation of $[M(CNPh)_6]^*$ occurs in chloroform solution prompted us to initiate a series of experiments aimed at the reduction of protons to hydrogen in aqueous solution. We decided to synthesize ML_6 complexes with L = 2-pyridylisocyanide to obtain the required water solubility (6). It was hoped that an inner sphere two-electron excited-state redox reaction of the type sketched in Figure 7 would result, yielding hydrogen and ML_6OH^+ as products. To date we have managed to prepare $Mo(CN$-2-$pyridyl)_6$ but have not characterized its thermal and photochemical reactivity. We have encountered some undesired side thermal reactions in acidic aqueous solution and will have to characterize these before starting any photochemical investigations. The WL_6 complex, that we have not yet prepared, may be a better system for our photochemical studies. Thus we are still confident that a reaction such as outlined in

Figure 7. Proposed reduction of protons in water molecules bound to the pyridine nitrogens in $[W(CN\text{-}2\text{-}pyridyl)_6]^*$ to yield H_2 and WL_6OH^+ products. This type of photoreaction has not been observed.

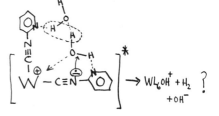

Figure 7 is possible, and we plan to continue our studies of such ML_6 systems in addition to the related work on the photochemistry of binuclear Rh(I) complexes that is described in the next section.

$Rh_2(bridge)_4{}^{2+}$

During investigations on the oligomerization of planar $Rh(CNR)_4{}^+$ complexes in solution (7, 8), we synthesized a bridged binuclear complex, $Rh_2(1,3\text{-diisocyanopropane})_4{}^{2+}$, that we now call rhodium bridge, Rh_2-$(bridge)_4{}^{2+}$ (9). It occurred to us that $Rh_2(bridge)_4{}^{2+}$ would be an interesting system to study in connection with multielectron redox processes, as each Rh^I could, in principle, furnish one or more electrons to a reducible substrate. Thus we have been engaged in a detailed study of both the thermal and the photochemical redox properties of this prototypal binuclear Rh(I) isocyanide complex. The rest of this chapter represents a progress report on this work.

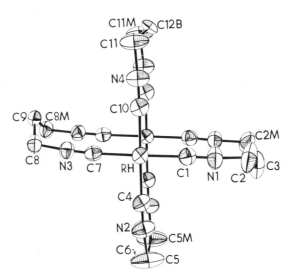

Figure 8. The $Rh_2(bridge)_4{}^{2+}$ structure

Now we have performed (*10*) an x-ray crystal structure analysis of [Rh$_2$(bridge)$_4$](BPh$_4$)$_2$. A view of the Rh$_2$(bridge)$_4^{2+}$ cation is shown in Figure 8. The binuclear complex has near D_{4h} symmetry, with a Rh–Rh distance of 3.26 Å. The occupied d_{z^2} orbitals on each d^8 planar Rh(I) center interact, yielding two MOs of symmetries a_{1g} and a_{2u}; the lowest unoccupied monomer orbitals (of a_{2u} symmetry) also interact and split into a_{1g} and a_{2u} levels in the binuclear complex. The orbitals of interest to us in discussing the low-lying absorption and emission bands and the photochemistry are, in order of increasing energy, $1a_{1g} < 1a_{2u} < 2a_{1g} < 2a_{2u}$. The ground state of Rh$_2$(bridge)$_4^{2+}$ is $^1A_{1g}$ ($1a_{1g}^2 1a_{2u}^2$).

The intense absorption band in the spectrum of Rh$_2$(bridge)$_4^{2+}$ at 553 nm (ϵ 14,500) in acetonitrile solution is attributed (*11*) to $^1A_{1g} \rightarrow {}^1A_{2u}$ ($1a_{2u} \rightarrow 2a_{1g}$), which is an allowed transition. The band falls well to the red of the analogous $^1A_{1g} \rightarrow {}^1A_{2u}$ ($d_{z^2} \rightarrow a_{2u}$) transition in a reference monomeric complex (e.g., this band in the spectrum of Rh(CNEt)$_4^+$ peaks at 380 nm) (*12*), which illustrates the importance of the axial orbital interactions (d_{z^2}-d_{z^2} and a_{2u}-a_{2u}) in the binuclear case.

Bright red emission is observed (*13*) on excitation of Rh$_2$(bridge)$_4^{2+}$ at 553 nm in acetonitrile solution. The emission peaks at 656 nm and the quantum yield is 0.056. The lifetime of the emission is very short ($\leqslant 2$ nsec), suggesting that the transition is spin-allowed, $^1A_{2u} \rightarrow {}^1A_{1g}$. Excitation of Rh$_2$(bridge)$_4^{2+}$ at 553 nm also gives rise to a relatively long-lived transient species ($\tau \sim 8$ μsec) that absorbs strongly between 400 and 500 nm. This transient is believed to be the $^3A_{2u}$ excited state of Rh$_2$(bridge)$_4^{2+}$ (*13*).

Irradiation of a blue 12M HCl solution of Rh$_2$(bridge)$_4^{2+}$ at 546 nm leads to clean conversion to a yellow product and H$_2$ (*11*). The absorption spectrum of the yellow product is identical with that of a sample of [Rh$_2$(bridge)$_4$Cl$_2$]$^{2+}$ prepared by chlorine oxidation of [Rh$_2$(bridge)$_4$]$^{2+}$ in 12M HCl solution. We have shown (*6, 11*) in several independent experiments that Rh$_2$(bridge)$_4^{2+}$ protonates in acidic aqueous solutions and that in HX (X = Cl, Br, I), this protonated complex binds X$^-$ as well. Thus we have formulated the photoreaction in 12M HCl as follows:

$$[\text{Rh}_2(\text{bridge})_4\text{H}]^{3+} \cdot \text{Cl}^- + \text{H}^+ + \text{Cl}^- \xrightarrow[12M \text{ HCl}]{546 \text{ nm}}$$
blue ($\lambda_{\max}$ 578 nm; ϵ 52,700)

$$[\text{Rh}_2(\text{bridge})_4\text{Cl}_2]^{2+} + \text{H}_2 \quad (4)$$
yellow ($\lambda_{\max}$ 338 nm; ϵ 56,200)

The photoreaction is uphill and proceeds in the presence of 1 atm of H$_2$ at 28°C. Under such conditions, the thermal back reaction between

$[Rh_2(bridge)_4Cl_2]^{2+}$ and H_2 to yield $[Rh_2(bridge)_4H]^{3+} \cdot Cl^-$ requires several days to go to completion. Our studies of the kinetics of the back reaction are not finished, but we do know that at high $[Cl^-]$, the reaction is first order in $[H_2]$ and inverse first order in $[Cl^-]$.

We are engaged now in studying the dependence of the quantum yield for the photooxidation of $[Rh_2(bridge)_4H]^{3+} \cdot X^-$ on the activities of H^+ and X^-. A sample of the results we have obtained is set out in Table II. It is apparent that the photoreaction strongly depends on both a_{H^+} and the nature of X^-. The relatively high quantum yield of 0.044, measured in degassed $9M$ HBr solution, shows that Br^- is more effective than Cl^- in promoting the photoreaction.

The experiments in HCl solutions establish that the quantum yield of the photoreaction increases sharply as a_{H^+} increases. A detailed analysis will have to await studies of the a_{H^+} dependence in media of constant

Table II. Quantum Yields for the Photooxidation of $Rh_2(bridge)_4{}^{2+}$ in HX Solutions at 29°C

$[HCl]$, M	$10^2 \Phi$ (degassed)[a]
12.8	0.79
12.1	0.56
11.1	0.26
10.1	0.083
9.1	0.028
8.1	0.010
1.0	too small to measure
$[HBr]$, M	
9	4.4

[a] Based on measurements of the appearance of $Rh_2(bridge)_4X_2{}^{2+}$.

a_{Cl^-}. What we can say at the present time is that the data at least are not inconsistent with a mechanism in which the key step involves attack of H_3O^+ on an excited $[Rh_2(bridge)_4H]^{3+} \cdot Cl^-$ species, producing hydrogen directly in a two-electron transfer process. We now are planning a series of experiments aimed at determining the lifetimes of the excited states and intermediates that are involved in this photoreaction.

Concluding Remarks

We have shown that it is possible to achieve an uphill reduction of protons to hydrogen by visible excitation of a binuclear Rh(I) complex. The charge transfer excited state involved in the photoreaction must be $^1A_{2u}$ or $^3A_{2u}$. In either case, the MO configuration is $(1a_{1g})^2(1a_{2u})^1(2a_{1g})^1$. If we assume, as seems reasonable, that the $2a_{1g}$ level possesses substantial

$1s(H)$ character, then the excited state can be formulated as $[(Cl^-)-Rh^{I\frac{1}{2}}-Rh^{I\frac{1}{2}}-(H\cdot)]^{2+*}$. Attack by H_3O^+ could induce a second electron transfer ($1a_{2u} \rightarrow H\cdot$), yielding hydrogen and the two-electron oxidative–addition product, $[Rh_2(bridge)_4Cl_2]^{2+}$:

$$[(Cl^-)Rh^I-Rh^I(H^+)]^{2+} \xrightarrow{546\ nm} [(Cl^-)-Rh^{I\frac{1}{2}}-Rh^{I\frac{1}{2}}-(H\cdot)]^{2+*}$$

$$[(Cl^-)-Rh^{I\frac{1}{2}}-Rh^{I\frac{1}{2}}-(H\cdot)]^{2+*} \xrightarrow[HCl(aq)]{fast} [Cl-Rh^{II}-Rh^{II}-Cl]^{2+} + H_2$$

It is attractive to propose that axial ligand–metal interactions may operate as suggested earlier for Nu: $\rightarrow$ [ML]* (*see* "Introduction"), thereby reducing the fraction of excited species returning to the Rh^I–Rh^I ground state by back electron transfer. The σ-bonding interactions along the internuclear axis Cl–Rh–Rh–H in an $(1a_{1g})^2(1a_{2u})^1(2a_{1g})^1$ excited state are predicted to be larger than those in the ground state $[(1a_{1g})^2-(1a_{2u})^2]$; therefore, these interactions favor the H_2-producing pathway ($1a_{2u} \rightarrow H\cdot$) over back transfer ($1a_{2u} \leftarrow H\cdot$), as the energy of the $1a_{2u}\sigma^*$ level increases sharply as the extent of Cl–Rh–Rh–H coupling increases.

There is a possibility that the photoreaction we have described could be used as part of a solar-driven water-splitting cycle, $H_2O \xrightarrow{h\nu} H_2 + 1/2\ O_2$. Completion of the cycle requires that $Rh_2(bridge)_4Cl_2^{2+}$ oxidize H_2O to $1/2\ O_2 + 2H^+$, either directly or indirectly in a photochemical or thermal reaction. Thus we are exploring, in some detail, the reaction properties of $Rh_2(bridge)_4Cl_2^{2+}$ in aqueous solution. An experiment in which $Rh_2(bridge)_4^{2+}$ is used to assist a solar-driven photoelectrolysis of water is being pursued in collaboration with Mark Wrighton of M.I.T. The idea of this experiment is to produce hydrogen by irradiation of a solution of $[Rh_2(bridge)_4H]^{3+} \cdot X^-$ in the cathode compartment and to produce oxygen by irradiation of a semiconductor photoanode (e.g., TiO_2) that will, at the same time, assist the conversion of $Rh_2(bridge)_4X_2^{2+}$ back to the active species at the cathode, $[Rh_2(bridge)_4H]^{3+} \cdot X^-$.

Acknowledgment

It is a pleasure to acknowledge the many helpful discussions we have had with George S. Hammond, Vince Miskowski, and Mark Wrighton during the course of this work. Our investigations have been supported by the National Science Foundation. Matthey–Bishop, Inc. is acknowledged for a generous loan of rhodium trichloride. This is Contribution No. 5665 from the Arthur Amos Noyes Laboratory.

Literature Cited

1. Malatesta, L., Sacco, A., Ghielmi, S., *Gazz. Chim. Ital.* (1952) **82**, 516.
2. Thich, J. A., unpublished data.
3. Mann, K. R., Ph.D. thesis, California Institute of Technology (1977).
4. Mann, K. R., Cimolino, M., Geoffroy, G. L., Hammond, G. S., Orio, A. A., Albertin, G., Gray, H. B., *Inorg. Chim. Acta* (1976) **16**, 97.
5. Mann, K. R., Hammond, G. S., Gray, H. B., *J. Am. Chem. Soc.* (1977) **99**, 306.
6. Lewis, N. S., M.S. thesis, California Institute of Technology (1977).
7. Mann, K. R., Gordon, J. G., II, Gray, H. B., *J. Am. Chem. Soc.* (1975) **97**, 3553.
8. Mann, K. R., Lewis, N. S., Williams, R. M., Gordon, J. G., II, Gray, H. B., *Inorg. Chem.*, in press.
9. Lewis, N. S., Mann, K. R., Gordon, J. G., II, Gray, H. B., *J. Am. Chem. Soc.* (1976) **98**, 7461.
10. Mann, K. R., Thich, J. A., Lewis, N. S., Gray, H. B., unpublished data.
11. Mann, K. R., Lewis, N. S., Miskowski, V. M., Erwin, D. K., Hammond, G. S., Gray, H. B., *J. Am. Chem. Soc.* (1977) **99**, 5525.
12. Isci, H., Mason, W. R., *Inorg. Chem.* (1975) **14**, 913.
13. Miskowski, V. M., Nobinger, G. L., Kliger, D. S., Hammond, G. S., Lewis, N. S., Mann, K. R., Gray, H. B., *J. Am. Chem. Soc.* (1978) **100**, 485.

RECEIVED September 20, 1977.

4

Photochemical and Photophysical Processes in 2,2'-Bipyridine Complexes of Iridium(III) and Ruthenium(II)

R. J. WATTS, J. S. HARRINGTON, and J. VAN HOUTEN

Department of Chemistry, University of California, Santa Barbara, CA 93106

> Tris bipyridyl complexes of Ru(II) and Ir(III) have been found to be luminescent and photochemically active in fluid solutions—hence the term "luminactive." Luminescence lifetimes and quantum yields for emission and photochemistry provide a direct measure of the competition between chemical and physical energy transformations in these species. Photoactivity of $Ru(bpy)_3{}^{2+}$ in aqueous HCl and aqueous $NaHCO_3$ is partially caused by photoanation. The primary photoproducts have been isolated and studied by absorption and emission spectroscopy. The results are consistent with their identification as species containing monodentate bpy. Photoactivity of $Ir(bpy)_3{}^{3+}$ in aqueous NaOH gives rise to $[Ir(bpy)_2OH(bpy)]^{2+}$. A model based on a double potential minimum is presented to interpret the luminescence and photochemical properties of $Ru(bpy)_3{}^{2+}$.

The production of electronic excited states in polyatomic molecules by absorption of visible and UV irradiation results in a variety of energy transfer processes. These include both radiative processes, which result in luminescence, and nonradiative processes. The latter can be conveniently divided into: (a) photophysical processes in which the chemical identity of the absorbing species is retained and the absorbed energy is eventually transferred as heat into the environment, and (b) photochemical processes in which the chemical identity of the absorbing species is altered through either endothermic or exothermic reactions of the excited states resulting from light absorption. These chemically reactive excited states can be reached either by direct absorption of light or indirectly by light absorption followed by energy transfer within the reactive species or via sensitization by a second absorbing species.

While it is convenient to divide nonradiative energy transfer into photophysical and photochemical processes, the classification of a given process is not always as clear as the division implies. Among the major complicating features encountered are: (a) how large a geometrical distortion constitutes a change in chemical identity, and (b) over what time increment following light absorption is the classification to be made?

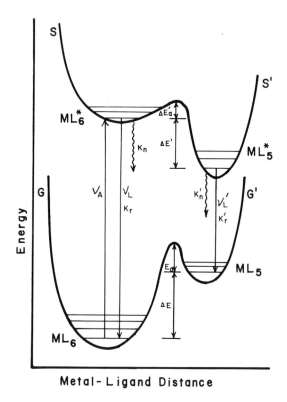

Figure 1. *Energy transfer in double minima potential wells*

Consider the following example, diagrammed in Figure 1, which serves to illustrate these features. The ML_6 molecule, which is a central metal bound to three bidentate ligands and which has overall trigonal geometry in the equilibrium ground state configuration, undergoes a Franck–Condon transition to an excited state having nearly the same equilibrium geometry along the metal–ligand bond axes. The potential well for this excited state (S) interacts with that for a second excited state (S') under distortion along the M–L axis. This second excited state has an equilibrium geometry in which the M–L distance is substantially increased because of migration of metal electron density from the pseudo-octahedral faces to a M–L bond axis. This distorted excited state can undergo a vertical transition to an equivalently distorted ground state via a radiative emission in which metal–electron density along the M–L axis migrates back to the pseudo-octahedral faces. In this example, the lowest-energy ground state configuration is considered to be the trigonally distorted ML_6 species whereas the lowest-energy electronically excited state is the $ML_5{}^*$ species.

The experimental observations one might make on such a system would depend critically upon environmental factors such as temperature, solvent, viscosity, etc. For example, at low temperatures ($kT << \Delta E_a' < \Delta E_a$) one would expect to see simply emission (k_r) or radiationless decay (k_n) of S as it returns to G, and classification of the energy transfer as photophysical would be straightforward. At higher temperatures ($\Delta E_a' \leq kT < \Delta E_a$) one might observe radiative (k_r') or radiationless (k_n') processes from S' to G' as well as from S to G, and the distorted G' isomer might be observable in a noncoordinating solvent medium. Again, assuming that G' could be observed and characterized as a chemical species distinct from G, classification of photochemical and photophysical events would be distinct, though complicated. Finally, consider the case where either $\Delta E_a' < \Delta E_a < kT$ or $\Delta E_a' \sim \Delta E_a < kT$; one might see emission from S' to G' as the only radiative process and might be unable to observe G' assuming ΔE and $\Delta E' >> kT$. Here classification of the energy transfer events is difficult since photochemistry may or may not have occurred as an intermediate step, depending on the extent of distortion present in G', although no net photochemistry would be observed.

Several points are apparent in this example. First, the dividing line between photochemistry and photophysics is indistinct since it depends on the extent of geometrical distortion between potential minima necessary to define a "distinct chemical species" as opposed to a "highly distorted excited state." Operationally, this distinction is complicated further by activation energies which may or may not enable the experimentalist to observe species corresponding to a potential minimum at a given temperature. Second and more important than the semantic difficulties in distinguishing photochemistry and photophysics is the point that isolated observations of a species under fixed environmental parameters can lead to vastly different observations whose relationship may not be immediately apparent. Only through extensive studies of both photochemistry and photophysics over a wide range of environmental parameters can the experimentalist hope to arrive at a cohesive interpretation of the myriad of energy transfer events subsequent to electronic excitation of a polyatomic species.

A variety of well-established techniques have been used in studies of photochemical and photophysical energy transfer processes resulting from electronic excitation of metal complexes (1, 2). Excited-state absorption techniques have proved to be extremely useful in studies of both photochemistry and photophysics, but the requirement for extremely high intensity, fast-pulsed light sources limits the scope of this method. Although luminescence often has been used to characterize photophysical energy transfer, these studies generally have been performed under conditions where photochemistry is absent (glasses or crystalline solids at 4.2°–77°K). By variations in temperature and/or solvent, one can hope to find that species which exhibit low-temperature luminescence begin to undergo photochemical activity while still displaying luminescence. Such species are referred to as "luminactive," and constitute a class of compounds around which the discussion will be centered. While the present range of luminactive compounds at temperatures above 273°K is very limited, there are undoubtedly many more compounds that are luminactive between 77° and 273°K. Hence, studies of luminactive species will become more common as photophysical studies are extended upward from 77°K and as photochemical studies are extended downward from 273°K.

The purpose of this discussion is to present an analysis of photophysical and photochemical energy-transfer events in two such luminactive species. At the present time, a comprehensive model of the energy transfer processes in $Ru(bpy)_3^{2+}$ can be presented. Far less work has been completed on $Ir(bpy)_3^{3+}$, but its luminactive nature has been established.

Experimntal

Luminescence and Absorption Measurements. Corrected luminescence spectra were determined with either a Perkin–Elmer Hitachi MPF-3A spectrophotofluorimeter (250–700 nm) or with a red-sensitive apparatus described previously (3) (450–1100 nm). The dependence of the response of each instrument on wavelength was determined by calibration against a standard NBS quartz–iodine lamp. Absorption spectra were determined with a Cary 15 spectrometer.

Photolysis of $Ru(bpy)_3^{2+}$ and $Ir(bpy)_3^{3+}$ Solutions. Solutions of $10^{-3}M$ trisbipyridylruthenium(II) chloride in $1M$ HCl or 0.01 MNaHCO$_3$ and $10^{-3}M$ trisbipyridyliridium(III) chloride in $0.01M$ NaOH were prepared and were stored in the dark at 273°K when not in use. For photolysis runs, samples of these solutions were placed in 1-cm rectangular borosilicate glass cells and were heated to 363°K in a thermostated cell holder with a Haake FK circulating water bath while dry nitrogen gas was bubbled through the solution to remove dissolved oxygen.

For photolysis of the $Ru(bpy)_3^{2+}$ solutions, the output of a 1000-W Hanovia Hg–Xe arc was passed through a 6-cm, water-cooled copper sulfate solution filter and was focused on the cell in the thermostated holder. The copper sulfate filter was removed for photolysis of $Ir(bpy)_3^{3+}$ solutions. Cells were capped tightly prior to photolysis to prevent diffusion of oxygen back into the nitrogen-saturated solutions. Under these conditions, photolysis times of 15–20 min at 363°K were generally used to obtain maximum product yields. Dark solutions heated to 363°K for periods of several hours were found to be thermally stable.

Following photolysis, the solutions were cooled quickly to ice temperature to minimize thermal reactions of the photoproduct. Purification of the product obtained by photolysis of Ru(bpy)$_3^{2+}$ in 0.01M NaHCO$_3$ was accomplished easily by passing the photolysis solution directly onto a 1 cm × 5 cm column of Sephadex CM cation exchange resin in 0.01M NaHCO$_3$. The deep-red photoproduct was eluted immediately from the column with water, leaving the orange-red starting material behind. To isolate the photoproduct from the solutions of Ru(bpy)$_3^{2+}$ photolyzed in 1M HCl, it was necessary to remove the solvent with a small freeze-drying apparatus. The red-purple residue was dissolved in methanol and was purified by column chromatography on a 2.5 cm × 50 cm column of Sephadex LH-20, using 0.01M HCl in methanol for elution. An orange band of unreacted Ru(bpy)$_3^{2+}$ eluted first from the column, followed by a band containing the red photoproduct. When the chromatography was carried out in room light, a third yellow band merged into the top of the red band. This band also was observed in increased proportion when photolysis times were extended beyond 20 min. For photolysis solutions irradiated for 15 min and chromatographed in the dark, this third band was not observed. Solutions from the photolysis of Ir(bpy)$_3^{3+}$ in 0.01M NaOH were studied without further purification by absorption and emission spectroscopy.

Solutions of purified Ru(bpy)$_3^{2+}$ photoproducts were studied within 1 hr of isolation when kept at room temperature because of their thermal instability. Solutions frozen at 77°K for low-temperature emission studies were found to be stable for a period of several days.

Preparation of Complexes. [Ru(bipy)$_3$]Cl$_2$ was purchased from G. F. Smith Chemical Co. and was used without further purification. [Ir(bpy)$_3$][NO$_3$]$_3$ was prepared by the method described by Flynn and Demas (4, 5), and [Ir(bpy)$_2$H$_2$O(bpy)]Cl$_3$ was prepared by the method described by Watts, Harrington, and Van Houten (6).

Results

Photolysis of Ru(bpy)$_3^{2+}$ in 0.01M NaHCO$_3$. Absorption and emission spectra obtained from solutions of the purified product of photolysis of Ru(bpy)$_3^{2+}$ in 0.01M NaHCO$_3$ are shown in Figure 2. The absorption spectrum of the purified photoproduct in 0.01M NaHCO$_3$ has intense bands at 510 and 355 nm. A structured emission with a maximum at 705 nm is observed in the luminescence of the compound when excited at 436 nm in 0.01M NaHCO$_3$ in water–methanol glasses at 77°K. When the pH of a purified solution of the photoproduct was adjusted to a value of 1 with HClO$_4$, the absorption spectrum was found to blue shift. As shown in Figure 2, absorption bands at 480 and 345 nm are observed at pH = 1, and the low-temperature emission maximum shifts to 660 nm. The absorp-

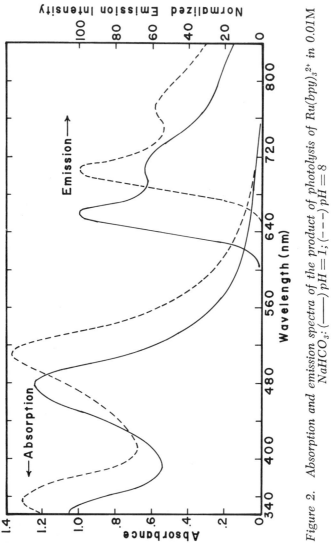

Figure 2. Absorption and emission spectra of the product of photolysis of $Ru(bpy)_3^{2+}$ in 0.01M $NaHCO_3$: (———) $pH = 1$; (– – –) $pH = 8$

tion bands at 510 and 355 nm could be regenerated by treatment of the $HClO_4$ solution with NaOH to adjust the pH to 10.

Photolysis of $Ru(bpy)_3^{2+}$ in 1M HCl. The absorption and low-temperature emission spectra of freshly purified solutions of the photoproduct are shown in Figure 3. Intense absorption bands at 495 and 350 nm are observed in the absorption spectrum and a weak, structured emission maximizing at 680 nm is observed in the low-temperature emission. Both absorption and emission spectra of the freshly prepared photoproduct were found to be independent of pH over the range 1–8 when 1M NaCl was used as a supporting electrolyte. In the absence of supporting NaCl, the 495-nm absorption of the photoproduct was found to be red-shifted to about 510 nm by adjustment of the pH to 8 with 0.01M

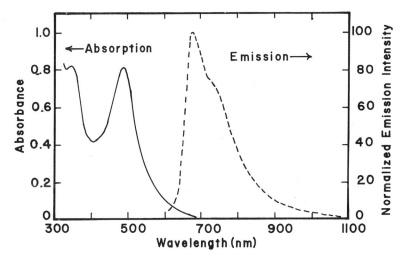

Figure 3. Absorption and emission spectra of the product of photolysis of $Ru(bpy)_3^{2+}$ in 1.0M HCl

$NaHCO_3$. Addition of $HClO_4$ to the pH = 8 solution to readjust the pH to 2 did not regenerate the original 495-nm absorption of the photoproduct, but led to a new absorption at about 480 nm. Addition of 1M NaCl to either the pH = 8 or pH = 3 solutions led to shifts in the respective absorption bands back to 495 nm. The 495-nm absorption band in aqueous solutions was shifted to 508 nm in methanol. Evaporation of the methanol followed by dissolution in 0.01M HCl with 1M NaCl supporting electrolyte regenerated the 495-nm absorption band. However, when the photoproduct was dissolved in acetonitrile two new bands appeared at 480 and 540 nm. Evaporation of the acetonitrile followed by partial dissolution of the residue in water gave a solution with an absorption band at 455 nm and an emission characteristic of $Ru(bipy)_3^{2+}$. The water insoluble por-

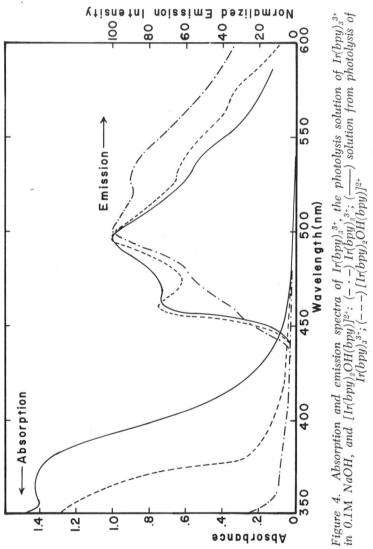

Figure 4. Absorption and emission spectra of $Ir(bpy)_3^{3+}$, the photolysis solution of $Ir(bpy)_3^{3+}$ in 0.1M NaOH, and $[Ir(bpy)_2OH(bpy)]^{2+}$: (–·–) $Ir(bpy)_3^{3+}$; (———) solution from photolysis of $Ir(bpy)_3^{3+}$; (– – –) $[Ir(bpy)_2OH(bpy)]^{2+}$

tion of the residue was redissolved in acetonitrile and was found to have an absorption band at 545 nm.

Photolysis of Ir(bpy)$_3^{3+}$ in 0.01M NaOH. Absorption and emission spectra of $10^{-3}M$ solutions of Ir(bpy)$_3^{3+}$ in 0.01M NaOH before and after photolysis at 363°K are shown in Figure 4. Absorption and emission spectra of [Ir(bpy)$_2$OH(bpy)]$^{2+}$ are included in the figure for comparison. The photolysis results in the growth of an absorption band at 370 nm. A redistribution of intensity occurs in the emission spectrum because of a relative growth of a band at 460 nm and a shift of the intense band at 500 nm in the starting material to 495 nm in the photolyzed solution.

Discussion

Previous studies of the luminescence lifetimes (7, 8), luminescence quantum yields (7), and photochemical quantum yields (9) of Ru(bpy)$_3^{2+}$ as a function of temperature over the range 278°–368°K in aqueous acid solutions clearly indicate that measurable photochemistry begins to occur at temperatures where luminescence quantum yields and lifetimes are also measurable (7). Hence, the term luminactive is applicable to this species. The interpretation of the combined photophysical and photochemical results requires a cohesive model which quantitatively fits the quantum yield and lifetime data and qualitatively correlates with the observed photoreactivity. Such a model is presented in Figure 5. In this model, k_{1r} represents the radiative decay rate of the charge-transfer manifold of emitting levels, and k_{1q} represents their rate of radiationless decay to the ground state. Photoreactivity as well as luminescence quenching is believed to be caused by a set of crystal field levels which lie above the emitting charge-transfer levels with an energy gap, ΔE_{LF}. These levels undergo radiationless decay back to the ground state with rate constant k_{2q} or undergo photochemistry with an additional activation energy, $\Delta E_a^{\ddagger}$, with rate constant k_p. For this model, the overall photochemical quantum yield as a function of temperature, $\Phi(T)$, is given by

$$\Phi(T) = \frac{k_{2q}e^{-\Delta E_{LF}/kT}}{k_{1r} + k_{1q} + k_{2q}e^{-\Delta E_{LF}/kT}} \cdot \frac{k_p e^{-\Delta E_a^{\ddagger}/kT}}{k_{2q} + k_p e^{-E_a^{\ddagger}/kT}} \quad (1)$$

The luminescence lifetime, $\tau_m(T)$, and luminescence quantum yield, $Q(T)$, are given, respectively, by

$$\tau_m(T) = [k_{1r} + k_{1q} + k_{2q}e^{-\Delta E_{LF}/kT}]^{-1} \quad (2)$$

$$Q(T) = k_{1r}\tau_m(T) \quad (3)$$

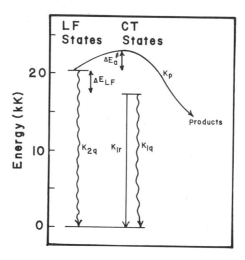

Figure 5. Model for energy transfer in $Ru(bpy)_3^{2+}$

By computer fitting of this model to experimental measurements of $\Phi(T)$, $\tau_m(T)$, and $Q(T)$ over the temperature range 278°–368°K, the set of rate constants and energy gaps shown in Table I have been obtained (7, 9). With these parameters the model provides a consistent, quantitative interpretation of the photophysical and photochemical information available for $Ru(bpy)_3^{2+}$ in aqueous solutions.

The photolysis results obtained in 1M HCl and in 0.01M NaHCO$_3$ clearly indicate that the nature of the photoproduct obtained is dependent upon the anion present in solution. Furthermore, previous studies have shown that the photochemical quantum yield is dependent upon the chloride ion concentration in HCl solutions, with higher quantum yields at higher Cl⁻ concentrations (9). Hence, we conclude that at least a portion of the photochemical activity is attributed to photoanation.

While most synthetic and kinetic studies of the binding of bpy to transition metal ions have assumed that species containing monodentate bpy are unstable with respect to either labilization or chelation of the bidentate ligand (10, 11), we have recently isolated a stable complex

Table I. Parameters Obtained from Measurements of Luminescence Lifetimes, Luminescence Quantum Yields, and Photochemical Quantum Yields of $Ru(bpy)_3^{2+}$ in Aqueous Solutions (7, 9)

Rate Constants (sec⁻¹)	Energy Gaps (kK)
$k_{1r} - 6.9 \times 10^4$	
$k_{1q} - 1.2 \times 10^6$	$\Delta E_{LF} - 3.6 \pm 0.1$
$k_{2q} - 1 \times 10^{13}$	$E_a^{\neq} - 2.0 \pm 0.1$
$k_p - 1.0 \times 10^{13}$	

containing monodentate bpy (6). This complex, which contains Ir(III) bound to two bidentate bpys, one monodentate bpy, and one water molecule (which can be converted to hydroxide with a pKa of 3), is thermally stable both in acid and base solutions and as a pure crystalline solid. In the present study, it is evident from comparison of the absorption and emission spectra of the solution obtained by photolysis of Ir(bpy)$_3^{3+}$ with the absorption and emission spectra of [Ir(bpy)$_2$OH-(bpy)]$^{2+}$ (see Figure 4) that the latter species is formed in the photolysis.

By analogy with the Ir(III) results, one might expect formation of a complex of Ru(II)-containing monodentate bpy in the photolysis of Ru(bpy)$_3^{2+}$. Consideration of the results we have obtained indicates they are indeed consistent with such a formulation. Photolysis in 1M HCl leads to formation of a species with absorption bands at 495 and 350 nm and an emission at 680 nm. The RuCl$_2$(bpy)$_2$ complex is characterized by absorption bands at 553 and 380 nm in acetonitrile (12, 13) and an emission at 715 nm (14, 15) whereas the Ru(bpy)$_3^{2+}$ starting material has an absorption band at 455 nm and a low-temperature emission maximum at 590 nm. Hence, the photoproduct in 1M HCl has absorption and emission features at energies intermediate between those of Ru(bpy)$_3^{2+}$ and RuCl$_2$(bpy)$_2$.

The long wavelength absorption and emission bands of Ru(bpy)$_3^{2+}$ and RuCl$_2$(bpy)$_2$ have been assigned as MLCT (12, 13, 14, 15) whereas the shorter-wavelength (380 nm) absorption band of the latter complex has been assigned as LMCT (12, 13). These trends clearly indicate red shifts in both the MLCT and LMCT bands of Ru(II) when π-donating chloride ligands are introduced into the coordination sphere. The intermediate nature of these bands in the photoproduct indicate that one rather than two chlorides is present in the coordination sphere. Furthermore, the absence of a pH dependence in the absorption of the photoproduct when 1M NaCl is present rules out the presence of water or hydroxide in the coordination sphere. Hence, our data are consistent with formulation of the photoproduct in 1M HCl as Ru(II) bound to two bidentate bipys, one monodentate bpy and one chloride, [Ru(bpy)$_2$Cl(bpy)]$^{1+}$.

By analogous reasoning, the product of the photolysis of Ru(bpy)$_3^{2+}$ in 0.01M NaHCO$_3$ is formulated as [Ru(bpy)$_2$OH(bpy)]$^{1+}$, which is converted to [Ru(bpy)$_2$H$_2$O(bpy)]$^{2+}$ by treatment with acid. This is consistent with red shifts in the absorption maxima to 510 and 355 nm when the photolysis is performed in 0.01M NaHCO$_3$ rather than 1M HCl, because of the superior π-donating ability of OH$^-$ relative to Cl$^-$. As expected, these absorption bands blue shift to 480 and 345 nm in acidic solutions because of the weaker π-donating ability of H$_2$O relative to either Cl$^-$ or OH$^-$.

From the dependence of the absorption spectra of the photoproducts on concentration of anions in solution, it appears that the coordinated Cl⁻ or OH⁻ is thermally labile in these species. Thus, the shift in the absorption of solutions containing [Ru(bpy)$_2$Cl(bpy)]$^{1+}$ from 495 nm in 1M NaCl to 510 nm in 0.01M NaHCO$_3$ with no NaCl added is attributed to conversion to [Ru(bpy)$_2$OH(bpy)]$^{1+}$ which can then be converted to [Ru(bpy)$_2$H$_2$O(bpy)]$^{2+}$ by acidification, as indicated above. While formation of Ru(bpy)$_3$²⁺ in purified solutions of the photoproduct from 1M HCl strongly supports our formulation of a species containing monodentate bpy, the peculiar effect of acetonitrile is not fully understood. Replacement of coordinated Cl⁻ by solvent is expected to occur in coordinating solvents on the basis of the thermal lability postulated above. Thus, [Ru(bpy)$_2$acn(bpy)]$^{2+}$ is expected to be formed in acetonitrile. Apparently this species is not stable with respect to chelation of the monodentate bpy to form Ru(bpy)$_3$²⁺. Identification of the second species formed by adding acetonitrile to [Ru(bpy)$_2$Cl(bpy)]$^{1+}$ will require further study to characterize fully.

We now turn our attention back to the luminescence properties of Ru(bpy)$_3$²⁺. The low-temperature and room-temperature luminescence has been widely studied, and both are reproduced in Figure 6 for reference. As this figure indicates, the low-temperature luminescence is highly structured and is widely accepted to be attributed to a set of MLCT excited levels. While it is tempting to assume that the room-temperature emission is caused by the same MLCT states, altered only by thermal redistribution of the relative population, the red shift and total absence of vibrational structure in the room-temperature luminescence suggest that the emission may have a substantially different origin than that at 77°K. Furthermore, previous studies have shown that although the effect of deuteration of the coordinated bpy on the luminescence lifetime is small, a large effect is seen when the solvent is changed from D$_2$O to H$_2$O (*3*). Also, extrapolation of the model used to analyze the low-temperature emission data (*16*) to room temperature leads to a predicted radiative rate constant of 12×10^4 sec^{-1} as compared with the measured value of 6.9×10^4 sec^{-1} (*7*). These factors suggest strong interactions of the excited state of Ru(bpy)$_3$²⁺ with water in fluid media, which we previously attributed to interactions of the MLCT configuration with charge-transfer-to solvent (CTTS) configurations (*3,7*). This by no means implies a complete electron transfer to solvent water to form aquated electrons but rather some penetration of the excited-electron density into the first solvation sphere in the fluid medium. It is interesting to note, however, that this quenching mechanism is in a sense analogous to the electron transfer mechanisms responsible for the quenching of the Ru(bpy)$_3$²⁺ luminescence by added electron acceptors. In fact, recent measurements indicate that two-photon processes can give small yields of solvated electrons under some circumstances (17), further stressing the potential analogy.

To continue this analogy, it also is known that energy transfer sometimes must occur to account for the quenching of excited $Ru(bpy)_3^{2+}$ by species that are not easily reduced or oxidized. By this analogy, it also is possible to explain the luminescence properties of $Ru(bpy)_3^{2+}$ in fluid

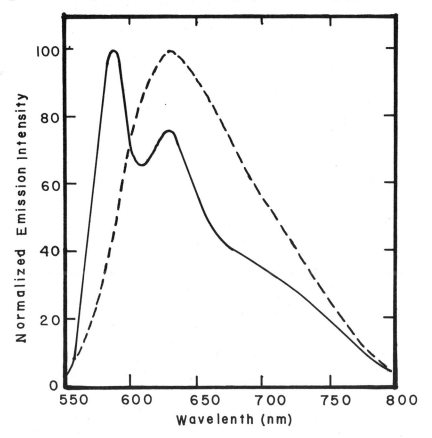

Figure 6. Emission spectra of $Ru(bpy)_3^{2+}$: (———) 77°K; (- - -) 273°K

water by mechanisms more closely associated with energy transfer than with electron tranfer. This alternate explanation is outlined in Figure 7, which is a slight modification of a diagram used by Dellinger and Kasha in their discussion of the use of a double-minima potential to interpret solvent effects on emission spectra (18). As applied to $Ru(bpy)_3^{2+}$, the potential well labeled by CT would represent the charge-transfer levels and that labeled by LF, the ligand field levels. At low temperature,

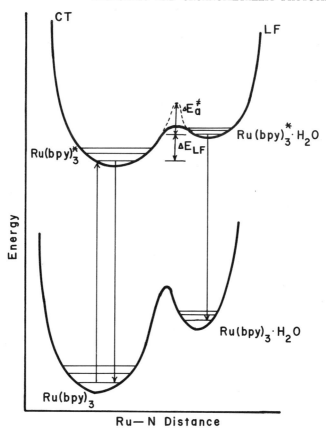

Figure 7. Double minima potential wells for $Ru(bpy)_3^{2+}$

emission would occur from the charge-transfer levels because of both thermodynamic (ΔE_{LF}) and kinetic ($\Delta E_a^{\ddagger}$) barriers to population of the ligand field levels. As indicated by Dellinger and Kasha, $\Delta E_a^{\ddagger}$ may include contributions from both solvent viscosity (dotted line) and intramolecular potential terms (solid line). In the case of $Ru(bpy)_3^{2+}$, the anticipated distortion along the Ru–N bond in the ligand field state is

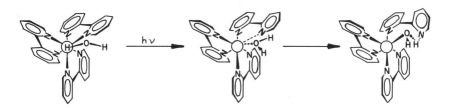

Figure 8. Proposed intermediate in the photolysis of $Ru(bpy)_3^{2+}$

likely to result in twisting of the plane of one of the pyridine rings about the C–C bridging bond of bpy. A viscous solvent would be expected to contribute a substantital potential barrier to this type of motion. In fluid media the solvent potential barrier would be greatly diminished, leading to a smaller value of $\Delta E_a^{\ddagger}$. As a result, an enhanced rate of population of the ligand field levels would be expected in fluid media. Furthermore, as the temperature is increased, higher equilibrium concentrations of the ligand field levels would ensue.

Since distortion along the Ru–N bond occurs in the ligand field levels, a solvent water molecule can occupy a position close to a coordination site when this state is formed. An intermediate structure of this type is illustrated in Figure 8, which depicts an associative reaction of $Ru(bpy)_3^{2+}$ with H_2O to form $[Ru(bpy)_2H_2O(bpy)]^{2+}$. Such an intermediate structure would be expected to show a substantial red shift in emission relative to the charge-transfer levels as indicated in Figure 7, and a large solvent deuterium effect is consistent with the close approach of water to a coordination site. Furthermore, the radiative rate constant of the ligand field levels would be expected to be smaller than that of the charge-transfer levels because of the inherent forbidden nature of $d–d$ transitions.

In conclusion, the emission of $Ru(bpy)_3^{2+}$ is consistent with a charge-transfer assignment in low-temperature rigid glasses and increasing contributions from a set of ligand field levels in fluid media at elevated temperatures. These levels also account for the photochemical activity observed at elevated temperatures. Whether this or the previously proposed CTTS model best explains the luminescence results is perhaps a moot point since it is difficult to envision experiments that could clearly distinguish the two. However, it is quite likely that the double-minimum potential model depicted in Figure 7 will be of widespread use in the interpretation of luminescence and photochemical data in luminactive systems.

Acknowledgment

Acknowledgment is made to the Committee on Research of the University of California, Santa Barbara, for support of this research.

Literature Cited

1. Balzani, V., Carassiti, V., "Photochemistry of Coordination Compounds," Academic, London, 1970.
2. Adamson, A. W., Fleischauer, P. D., "Concepts of Inorganic Photochemistry," Wiley, New York, 1975.
3. Van Houten, J., Watts, R. J., *J. Am. Chem. Soc.* (1975) **97**, 3843.
4. Flynn, C. M., Demas, J. N., *J. Am. Chem. Soc.* (1974) **96**, 1959.
5. Ibid. (1975) **97**, 1988.

6. Watts, R. J., Harrington, J. S., Van Houten, J., *J. Am. Chem. Soc.* (1977) **99**, 2179.
7. Van Houten, J., Watts, R. J., *J. Am. Chem. Soc.* (1976) **98**, 4853.
8. Allsopp, S. R., Cox, A., Jenkins, S. H., Kemp, T. J., Tunstall, S. M., *Chem. Phys. Lett.* (1976) **43**, 135.
9. Van Houten, J., Watts, R. J., *J. Am. Chem. Soc.*, in press.
10. Maestri, M., Bolletta, F., Serpone, N., Moggi, L., Balzani, V., *Inorg. Chem.* (1976) **15**, 2048.
11. Basolo, F., Pearson, R. G., "Mechanisms of Inorganic Reactions," 2nd ed., Wiley, New York, 1967.
12. Braddock, J. N., Meyer, T. J., *Inorg. Chem.* (1973) **12**, 723.
13. Weaver, T. R., et al., *J. Am. Chem. Soc.* (1975) **97**, 3039.
14. Klassen, D. M., Crosby, G. A., *Chem. Phys. Lett.* (1967) **1**, 127.
15. Klassen, D. M., Crosby, G. A., *J. Chem. Phys.* (1968) **48**, 1853.
16. Harrigan, R. W., Hager, G. D., Crosby, G. A., *Chem. Phys. Lett.* (1973) **21**, 487.
17. Meisel, D., Matheson, M. S., Mulac, W. A., Rabani, J., *J. Phys. Chem.* (1977) **81**, 1449.
18. Dellinger, B., Kasha, M., *Chem. Phys. Lett.* (1976) **38**, 9.

RECEIVED September 20, 1977.

Roles of Charge Transfer States in the Photochemistry of Ruthenium(II) Ammine Complexes

PETER C. FORD

Department of Chemistry, University of California, Santa Barbara, CA 93106

> Ru(II) complexes display a variety of electronic excited states and the chemistry of these is very rich. Here, our recent work with the Ru(II) ammine complexes will be summarized with particular emphasis placed on the solution photochemistry and spectroscopy of the pentaammine complexes $Ru(NH_3)_5L^{2+}$. The electronic spectrum of $Ru(NH_3)_6^{2+}$ is shown to display both ligand field and charge transfer to solvent absorptions. Irradiation of the former principally leads to ammonia substitution while that of the latter mostly leads to oxidation of Ru(II) to Ru(III), with stoichiometric formation of H_2. When L is an unsaturated ligand such as a substituted pyridine, the visible spectrum is dominated by metal-to-ligand charge transfer absorptions. However, the presence of spectrally unseen ligand field states appear to dictate the major photoreaction pathway (ligand substitution) in many cases, although the MLCT state is capable of undergoing inner sphere electron transfer to another metal center when L contains a second coordination site (e.g., pyrazine).

Interest in the photochemistry of Ru(II) has been considerable in recent years (1, 2, 3, 4) and has been spurred by the discoveries that excited states of certain aromatic nitrogen, heterocycle Ru(II) complexes can undergo either energy transfer (5) or electron transfer (6) with the appropriate substrates. In our laboratories, attention has been focused largely on understanding the solution spectroscopy and photochemistry of the pentaammine complexes $Ru(NH_3)_5L^{2+}$, where L can be either a saturated ligand such as NH_3 or H_2O or a π-unsaturated ligand such as an organonitrile or a substituted pyridine (7–18). The spectra of these species are markedly dependent on the nature of the unique ligand (19), and the photochemistries that these complexes display in fluid solution include pathways that can be assigned logically as characteristic to particular excited state types. Thus, models for the photochemical reactions of various $Ru(NH_3)_5L^{2+}$ complexes can be described, and it is on the basis of one model that we demonstrated that judicious variation of ligand substituents or of the solvent medium can tune photochemical properties of particular systems (10). Applications of such models and the understanding of the consequences of molecular perturbations on excited state properties are essential in the design of chemical systems for applications as the conversion of radiant energy to chemical potential energy.

Spectra and Types of Excited States

Table I summarizes the aqueous solution absorption spectra of $Ru(NH_3)_6^{2+}$, $Ru(NH_3)_5py^{2+}$, and $Ru(NH_3)_5(CH_3CN)^{2+}$, respectively. The spectrum of $Ru(NH_3)_6^{2+}$ (Figure 1) is typical of that seen for other saturated ammine Ru(II) complexes, and the spectra of $Ru(NH_3)_5H_2O^{2+}$

Table I. Comparison of Spectra of $M(NH_3)_5L^{n+}$ Complexes in Aqueous Solution

Complex	λ_{max} (nm)	$(M^{-1}\,cm^{-1})$	Assignment	Ref.
Ru(II)				
$Ru(NH_3)_6^{2+}$	390	35	LF	18
	275	640	CTTS	
$Ru(NH_3)_5py^{2+}$	407	7,770	MLCT	9
	244	5,090	IL	
$Ru(NH_3)_5(CH_3CN)^{2+}$	350	163	LF	13
	226	15,400	MLCT	
Rh(III)				
$Rh(NH_3)_6^{3+}$	305	134	LF	32
	255	101	LF	
$Rh(NH_3)_5py^{3+}$	302	170	LF	33
	259	2,930	IL	
$Rh(NH_3)_5(CH_3CN)^{3+}$	301	158	LF	33
	253	126	LF	
Co(III)				
$Co(NH_3)_6^{3+}$	472	56	LF	32
	338	46	LF	
$Co(NH_3)_5py^{+3}$	474	64	LF	34
	340	54	LF	
	252	3,160	IL	
$Co(NH_3)_5CH_3CN^{3+}$	467	57	LF	24
	333	54	LF	

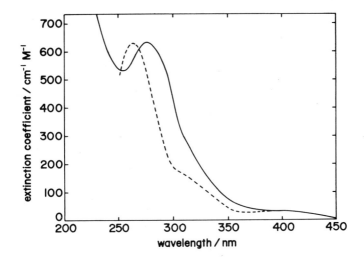

Figure 1. Electronic spectrum of $Ru(NH_3)_6^{2+}$ in aqueous solution (———) and in 80% aqueous ethanol (- - -); 25°C

and $Ru(en)_3^{2+}$ are qualitatively very similar. The lower energy, low intensity shoulder at 390 nm can be attributed to the $^1A_{1g} \rightarrow {}^1T_{1g}$ d-d transition predicted by ligand field (LF) theory. However, the maximum at 275 nm, although situated approximately where the predicted $^1A_{1g} \rightarrow {}^1T_{2g}$ LF band might be expected, has an extinction coefficient much larger than that seen for the analogous band of the Co(III) and Rh(III) homologues (see Table I). Given this observation, some preliminary photochemical results, and the relative ease of oxidizing $Ru(NH_3)_6^{2+}$ ($E_{\frac{1}{2}} = 0.05V$ for the $Ru(NH_3)_6^{3+/2+}$ couple (18)), we have suggested previously a charge transfer to solvent (CTTS) assignment (12, 15) (vide infra).

Replacement of one NH_3 by a π-unsaturated ligand leads to dramatic spectral changes as shown in the spectrum of $Ru(NH_3)_5py^{2+}$ (Figure 2). Thus the bands seen for $Ru(NH_3)_6^{2+}$ are obscured by the far more intense absorptions centered at 407 and 244 nm. These have been assigned respectively as a metal-to-ligand (py) charge transfer (MLCT) band and an internal ligand (IL)π-π* transition (19). The MLCT band position is markedly dependent on the nature of pyridine ring substituents, especially those in the para position, and on the nature of the solution medium (19). It is this sensitivity that forms the basis of the photochemical tuning mentioned above. Similar MLCT bands are seen for other unsaturated ligands such as the organonitriles. The LF band, although unseen in Figure 2, no doubt contributes in a minor way to absorption around 400 nm, given the similarity in the LF strengths of ammonia and pyridine. An illustration of this is seen in the spectrum of aqueous $Ru(NH_3)_5(CH_3CN)^{2+}$ (see Table I) that displays the expected lower energy LF band at 350 nm as a shoulder on the higher energy MLCT absorption centered at 226 nm.

Thus, there are four types of electronic transitions (CTTS, MLCT, IL, and LF) which should draw one's attention when evaluating the photochemistry of pentaammineruthenium(II) complexes. (Ligand-to-metal charge transfers unlikely are given the difficulty of reducing aqueous $Ru(NH_3)_6^{2+}$ (20).) On a rather qualitative level, one can examine the formal electronic character of these excited states and speculate regarding the reactivity to be expected for each. For example, a LF excited state involves an angular redistribution of charge primarily

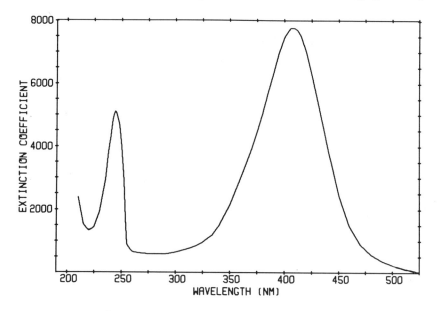

Figure 2. Electronic spectrum of $Ru(NH_3)_5py^{2+}$ in aqueous solution

located on the metal [i.e., $(NH_3)_5(Ru^*)L^{2+}$], and for the low spin d^6 complexes, such a state involves promotion of a t_{2g} electron to an e_g orbital, σ^* with regard to ligand–metal bonds. Thus, it is not surprising that ligand labilization is the most general reaction mode attributed to such states. In contrast, the IL excited state involves electronic redistribution localized on the ligand [$(NH_3)_5Ru(L^*)^{2+}$], and reaction patterns might involve ligand structural changes, reactions with other substrates, etc., similar to those of the free ligands. This is a relatively unexplored area of transition metal photochemistry, but examples such as certain photoreactions of coordinated stilbazole have been attributed to ligand-localized states (21, 22). A CTTS state can be represented formally as having an oxidized metal and a solvated electron [$(NH_3)_5$-RuL^{3+} ... e_s^-], and reduction of solvent or other solution substrate by the e_s^- is the reactivity most strongly suggested. Reactions characteristic of an oxidized metal center might occur also. Lastly, representation of the MLCT state as an oxidized metal coordinated to a radical ion [$(NH_3)_5(Ru^{III})(L^{\cdot})$]$^{2+}$ suggests reactivity such as electron transfer from another substrate to the oxidized metal center (23) or reactivity characteristic of the radical ligand. In the latter category, there might be electron transfer to another substrate (6, 14), structural rearrangement of the ligand (21, 22), or bimolecular reactions, such as electrophilic substitution on an aromatic $L^{\cdot}$ (9). Notably, for MLCT states of various Ru(II), all of the above reactivity types have been reported.

Complicating such simple considerations are the possibilities of the interaction and mixing of various excited state types, internal conversion from states initially populated to states of different multiplicities and/or different orbital parentages, and intermolecular energy transfer and quenching. Also, isoenergic crossing into the ground electronic state would give a highly excited vibrational state with considerably more energy than the activation energies of most thermal reactions seen for that complex or molecule. In case of such a mechanism, however, it appears that the reaction trajectory taken by such a hot ground state still reflects the nature of the electronic excited state from which it was derived (24).

$Ru(NH_3)_6^{2+}$

As noted above, preliminary results (7, 13) suggested the shorter wavelength band of the $Ru(NH_3)_6^{2+}$ spectrum to have CTTS character. Among these was the observation that the UV photolysis of several aqueous $Ru(NH_3)_5L^{2+}$ (L = N_2, CH_3CN, py, NH_3, or H_2O) gives Ru(III) products. The production of molecular hydrogen also was noted for some of these cases. In contrast, UV photolysis of the isoelectric $Rh(NH_3)_6^{3+}$ gives aquation products only. CTTS absorptions have been noted to be dependent on the nature of the solvent medium (25), and Figure 1 illustrates the difference between the $Ru(NH_3)_6^{2+}$ spectrum in aqueous solution and in 80% (v/v) aqueous ethanol. Notably, the lower energy band is unchanged, either in extinction coefficient or in position, consistent with its LF assignment. However, there is a shift of the shorter wavelength band to 264 nm, and as a consequence there is much better definition in 80% ethanol of a relatively weak shoulder at 310 nm, barely suggested in the aqueous spectrum. The shoulder very likely represents the $^1A_{1g} \to {}^1T_{2g}$ LF absorption predicted for this octahedral d^6 complex. (On the basis of this assignment, calculations using the approximate relationship $C = 4B$ for the Racah parameters give crystal field terms $\Delta = 26.8$ kK, $B = 454$ cm^{-1}, $C = 1816$ cm^{-1}.)

CTTS bands have been assigned previously on the basis of their linear energy correlation with the solvent dependence of the CTTS band of iodide (25, 26). This correlation is not seen for the charge transfer band of Ru(NH$_3$)$_6^{2+}$ (18), an unsurprising observation given that previous successes were seen only for anionic complexes. However, linear dependence of the energy of the charge transfer band with the mole fraction of water in various mixed solvents is seen (Figure 3); this suggests that solvation of the cationic Ru(II) and Ru(III) species is the dominant factor determining the shift of the CTTS band in different solvent environments.

Photolysis of acidic (pH ~ 3) aqueous Ru(NH$_3$)$_6^{2+}$ gives several products but analysis of these indicates two primary photoreactions: ammonia aquation (Reaction 1)

$$\text{Ru(NH}_3)_6^{2+} + \text{H}_2\text{O} \xrightarrow{h\nu} \text{Ru(NH}_3)_5\text{H}_2\text{O}^{2+} + \text{NH}_3 \qquad (1)$$

and oxidation of Ru(II) to Ru(III) (Reaction 2)

$$\text{Ru(NH}_3)_6^{2+} \xrightarrow{h\nu} \text{Ru(NH}_3)_6^{3+} \qquad (2)$$

Quantum yields measured over various irradiation wavelengths (λ_{irr}), 405–214 nm (Figure 4), show that photoaquation dominates (as expected)

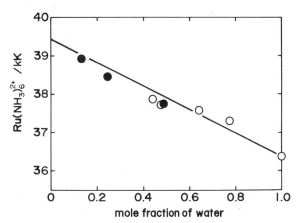

Figure 3. Plot of ν_{max} for the higher energy absorption band of Ru(NH$_3$)$_6^{2+}$ vs. the mole fraction of water in various mixed solvents; 25°C. (●) co-solvent is acetonitrile, (○) cosolvent is methanol or ethanol

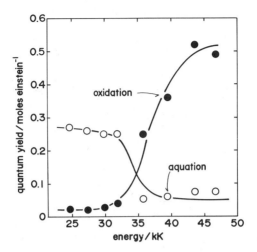

Figure 4. Variation of $\Phi_{Ru(III)}$ and Φ_{aq} with irradiation energy for $Ru(NH_3)_6{}^{2+}$ in aqueous solution

in the lower energy region (24.7–31.9 kK) where the LF bands make up most of the absorption but that photooxidation dominates with increasingly larger quantum yields at higher energies. The sharp increase in $\Phi_{Ru(III)}$ and the related decrease in Φ_{aq} correspond to the onset of the CTTS band.

At $\lambda_{irr} \geq 313$ nm (31.9 kK), both $\Phi_{Ru(III)}$ (0.03 ± 0.01 mol/einstein) and Φ_{aq} (0.26 ± 0.01) are essentially wavelength independent, suggesting the population of a common excited state as the result of irradiation in this region. A reasonable possibility is that this common state is the lowest energy LF singlet $^1T_{1g}$ and that the two reaction pathways represent competitive first-order deactivation pathways from this state. A reasonable scenario would be: (1) direct excitation into the LF singlet states; (2) relaxation to a common state, presumably $^1T_{1g}$, and (3) aquation via intersystem crossing into the LF triplet states competitive with oxidation via internal conversion from $^1T_{1g}$ into the CTTS state(s). This scenario is illustrated in Figure 5. Irradiation of the longer wavelength band of $Ru(NH_3)_5CH_3CN^{2+}$ (*see* Table I) gives only aquation products ($\Phi_{aq} = 0.26 \pm 0.03$, $\Phi_{Ru(III)}$ estimated to be $< 10^{-3}$ mol/einstein) (*13*). However, this result is consistent with the greater difficulty in oxidizing $Ru(NH_3)_5(CH_3CN)^{2+}$ ($E_{\frac{1}{2}} = 0.43$V for the Ru(III)/Ru(II) couple (*18*)), thus leading to a larger energy gap between the $^1T_{1g}$ and CTTS states.

For $\lambda_{irr} < 280$ nm, light absorption is directly into the CTTS state, and oxidation is dominant. Residual photoaquation also is seen; this feature may be attributed either to interconversion from the CTTS states

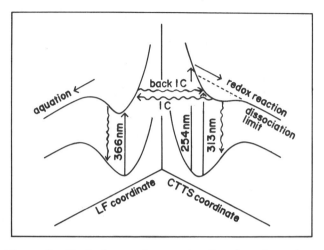

Figure 5. Excited state diagram for the photochemistry of aqueous $Ru(NH_3)_6^{2+}$. The aquation and redox pathways are shown on separate coordinates defined by the nature of the relevant excited states.

into the LF states or alternatively to direct absorption into LF bands obscured by more intense CT bands but still comprising a significant fraction of the absorption in that region. Hydrogen evolution was found to accompany the oxidation of the ruthenium complex in this region, and volumetric measurements showed the $Ru(III)/H_2$ ratio to be 2.0 ± 0.3. This result supports a reaction scheme such as:

$$Ru(II) \underset{CTTS^*}{\overset{h\nu}{\rightleftarrows}} (Ru(III) \ldots e_s^-) \qquad (3)$$

$$CTTS^* + H^+ \rightarrow Ru(III) + H\cdot \qquad (4)$$

$$H\cdot + H^+ \rightleftarrows H_2^+ \qquad (5)$$

$$\underline{H_2^+ + Ru(II) \rightarrow H_2 + Ru(III) \qquad (6)}$$

$$2Ru(II) + 2H^+ \rightarrow H_2 + 2Ru(III) \qquad (7)$$

This scheme is supported by the observations that $\Phi_{Ru(III)}$ is increased approximately twofold by increasing $[H^+]$ from $0.001M$ to $0.01M$ (*18*), indicating competition between trapping of CTTS* by H^+ (Reaction 4) and the relaxation of CTTS* back to starting material. Also, addition of 2-propanol to the reaction solutions led simultaneously to large increases in Φ_{H_2} and decreases in $\Phi_{Ru(III)}$ (*20*). This is consistent with the trapping of H· by the alcohol and the subsequent reaction of any Ru(III) products with the reducing radical $CH_3\dot{C}(OH)CH_3$. The current data do not differentiate, however, among the possible intermediacies of the oxidizing species H_2^+ or other oxidants possibly generated by the reaction of H· or H_2^+ with the chloride ($\sim 0.2M$ NaCl) used in the reaction solutions in maintaining ionic strength.

The excited state diagram shown in Figure 5 represents the CTTS state as having a discrete energy; however, it must be rather diffuse given the uncertainty and variety of the many solvent/complex configurations in the excited state. The very nature of CTTS states, especially for cationic complexes, remains a controversial topic. Solvated electrons generally have not been observed in the flash photolysis of cations (26), and attempts, in our laboratory, using a conventional xenon flash apparatus to study aqueous $Ru(NH_3)_6^{2+}$ were unsuccessful in this regard This certainly was no surprise given the rate constants reported for trapping of e^-_{aq} by $Ru(NH_3)_6^{3+}(7 \times 10^{10}$ M^{-1} sec^{-1} (27)) and by $H^+_{aq}(2.2 \times 10^{10}$ M^{-1} sec^{-1} (27)). The question remains whether the CTTS state represented in Reactions 3 and 4 actually involves the formation of a solvated electron or one more tightly bound to the Ru(III) center.

$Ru(NH_3)_5(py\text{-}x)^{2+}$ $(py\text{-}x = $ substituted pyridine$)$

Photolysis of the pyridine complex $Ru(NH_3)_5py^{2+}$ in aqueous solution at wavelengths shorter than 334 nm gives both photoaquation and photooxidation in a manner similar to other pentaammineruthenium(II) complexes (12). Although the major absorption band in this region is the IL $\pi\pi^*$ of the coordinated pyridine (see Figure 2), the principal reactions apparently occur from other states, perhaps produced by internal conversion/intersystem crossing from the initially populated IL configurations. At longer λ_{irr}, no photooxidation was detected, only photoaquation (Reaction 8) and a low quantum yield ($\sim 10^{-4}$ mol/ein-

$$Ru(NH_3)_5py^{2+} + H_2O \xrightarrow{h\nu} \begin{cases} (a) \; Ru(NH_3)_5H_2O^{2+} + py \\ (b) \; cis\text{-}Ru(NH_3)_4(H_2O)py^{2+} + NH_3 \\ (c) \; trans\text{-}Ru(NH_3)_4(H_2O)py^{2+} + NH_3 \end{cases} \quad (8)$$

stein) exchange of pyridine hydrogens with solvent hydrogen. The latter reaction is consistent with the formulation of the initially populated MLCT state as:

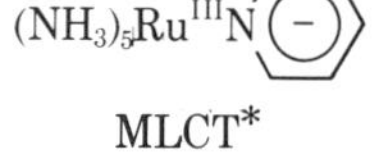

$$(NH_3)_5Ru^{III}N\!\!\!\bigcirc\!\!\!\ominus$$

MLCT*

The electron-rich coordinated pyridine of the excited state should be more susceptible to electrophilic attack by H^+ than when in the ground state, a prediction consistent with the observed photocatalysis of the hydrogen exchange. However, the Ru(III) center is generally substitution inert, especially toward ammonia aquation; thus the major photoreaction pathway (Reaction 8) is an unlikely consequence of the MLCT configuration. Accordingly, we proposed an LF state, populated by internal conversion from the initially formed MLCT state, as the direct precursor of ligand labilization (9, 10, 15).

Figure 6 shows a simple excited state model consistent with the above proposals for the pyridine complex. Manifolds of the MLCT and LF states are represented in recognition of the number of singlet and triplet states of each type predicted, especially when such considerations as spin-orbit coupling are made. To a first-order approximation, mixing of the MLCT and LF states is not allowed; therefore, the two excited state manifolds have differentiable characters described by their respective orbital parentages. Since quantum yields for the photoaquation of $Ru(NH_3)_5py^{2+}$ ($\Phi_{TOT} = \Phi_{NH_3} + \Phi_{py}$) are essentially independent of λ_{irr}, the model proposes that initial excitation is followed by efficient internal conversion to the lowest state, an LF state (Figure 6a) from which ligand aquation is one deactivation pathway.

Since the LF states are much less affected by the pyridine substituents (15) than are the MLCT states, choice of an appropriately electron-withdrawing x group should give the situation shown in Figure 6b, a lowest energy MLCT state. For such a complex, any ligand aquation noted would be predicted to come from the higher energy LF states. Thus, if deactivation to the lowest energy states is relatively efficient, complexes with lowest energy MLCT states should be much less substitution reactive than those with lowest energy LF states.

The sensitivity of the MLCT band to pyridine substituents allows one to use substituent effects to test these proposals. In Figure 7, Φ_{TOT} measured for irradiation at wavelengths corresponding to the charge transfer maximum $\lambda_{max}(MLCT)$ is plotted vs. $\lambda_{max}(MLCT)$. Notably, complexes with shorter wavelength MLCT bands are relatively photoactive with $\Phi_{TOT} \sim 0.1$ mol/einstein. However, all of the complexes with $\lambda_{max}(MLCT)$ longer than ~ 460 nm are significantly less active when irradiated at their $\lambda_{max}(CT)$, with Φ_{TOT} values as much as three orders of magnitude smaller. This pattern conforms to the excited state models of Figure 6 and suggests that the crossover point between complexes with a lowest energy LF state and those with a lowest energy MLCT state come when $\lambda_{max}(MLCT)$ is ~ 460 nm.

A key, unknown detail of the model would be the approximate energy of the lowest LF state of these complexes which, incidentally, is

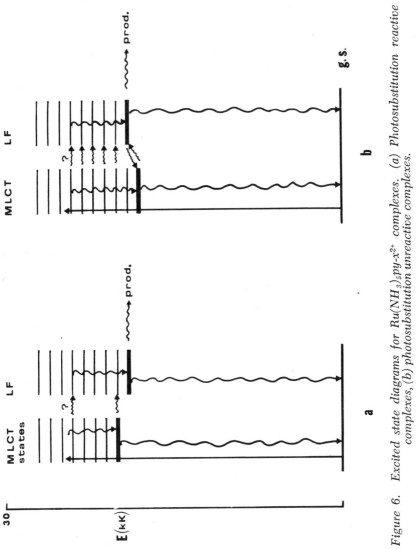

Figure 6. Excited state diagrams for $Ru(NH_3)_5py\text{-}x^{2+}$ complexes. (a) Photosubstitution reactive complexes, (b) photosubstitution unreactive complexes.

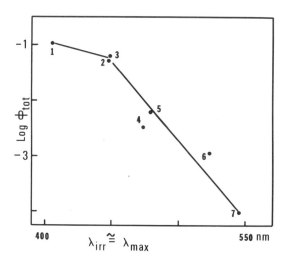

Figure 7. Φ_{TOT} for the photoaquation reactions of $Ru(NH_3)_5L^{2+}$ as a function of λ_{max} (MLCT) ($\lambda_{irr} \cong \lambda_{max}$ (MLCT)). L = (1) pyridine, (2) 4-phenylpyridine, (3) 3,5-dichloropyridine, (4) pyrazine, (5) isonicotinamide, (6) 4-acetylpyridine, (7) 4-formylpyridine.

assumed to be relatively independent of the ligand substituent. The MLCT bands are quite broad. For example, a point on the long wavelength tail corresponding to 5% of the absorbance at λ_{max} occurs at ~4 kK lower energy than at λ_{max}(CT). If this corresponds to the 0-0 absorption into the lowest MLCT (28), then the lowest MLCT state when $\lambda_{max} = 460$ nm (21.7 kK) would have an energy of ~17.7 kK. Since, at the crossover between the situations represented in Figures 6a and 6b, E(MLCT) should equal E(LF), this latter value should be a reasonable estimate of the lowest energy LF state. A similar energy is estimated for the $^3T_{1g}$ thexi state from the crystal field parameters of $Ru(NH_3)_6^{2+}$ (vide supra). However, attempts to evaluate such energies more directly by emission spectroscopy have been ambiguous (29). We have seen weak, LF-type emissions for several complexes with maxima at approximately 12-13 kK. A reasonable estimate of the Stokes shift would be 4 kK, thus placing the estimated LF thexi state energy at ~16-17 kK. Within the broad uncertainties of these methods, the three estimates give roughly comparable values for the energy (~ 16-18 kK) of the lowest LF state.

Another experimental observation differentiating those complexes with lowest energy LF states from those with lowest energy MLCT states is the dependence of Φ_{TOT} on λ_{irr} (Figure 8). The relatively reactive complexes, such as those of pyridine (λ_{max} 407 nm),3,5-dichloropyridine

(447 nm), and of 4-phenylpyridine (446 nm), show little dependence on λ_{irr}, consistent with a mechanism where most of the excitation leads to population of a substitution-reactive, lowest energy LF state. In contrast, the less reactive complexes with longer wavelength MLCT absorption bands, such as those of isonicotinamide (479 nm) and 4-acetylpyridine (523 nm), are markedly wavelength dependent, consistent with the view that initial MLCT excitation is followed by deactivation to the lowest MLCT*, with competitive crossing into the LF* manifold.

The $Ru(NH_3)_5py$-x^{2+} complexes display another remarkable example of excited state tuning, namely that by solvent effects. Although LF bands generally are insensitive to changes in the solvent (vide supra), MLCT absorption bands are markedly solvent dependent (*19*). For example, λ_{max}(MLCT) of the pyridine complex in DMSO appears ~40 nm to the red of its value in water. Thus, if a complex is borderline but reactive in water (Figure 6a), dissolving it in DMSO may reverse the order of its lowest energy states to give a situation conforming to Figure 6b.

Figure 9 demonstrates this effect. Although the patterns of behavior for the pyridine (λ_{max} 407 nm in H_2O and 447 nm in DMSO) and isonicotinamide (479 nm in H_2O and 511 nm in DMSO) complexes remain consistent (reactive and unreactive, respectively) in both solvents, the 4-phenylpyridine (446 nm in H_2O and 497 nm in DMSO) and 3,5-dichloropyridine (447 nm in H_2O and 491 nm in DMSO) complexes undergo behavioral changes. In water, both behave much like the reactive pyridine complex (relatively large, λ_{irr} independent quantum

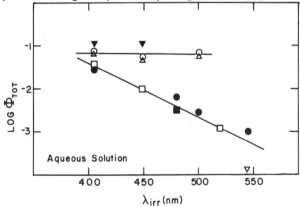

Figure 8. Wavelength dependence of log Φ_{TOT} for the photolysis of $Ru(NH_3)_5L^{2+}$ in aqueous solution. Symbols for various L's: (▼) pyridine; (○) 3,5-dichloropyridine; (△) 4-phenylpyridine; (▽) methylpyrazinium; (●) isonicotinamide; (■) pyrazine; (□) 4-acetylpyridine. Lines drawn for illustrative purposes only.

yields); in DMSO, their patterns are more similar to that of the unreactive isonicotinamide complex (relatively small, λ_{irr} dependent quantum yields).

Behavioral changes explainable by this model also have been seen for the red-shifting DMF solvent and for blue-shifting acetonitrile solvent (11). Thus, for those complexes where the lowest MLCT and LF states are reasonably close in energy, it appears possible to reverse the order of the lowest states and to change the general reactivity patterns by solvent variation alone.

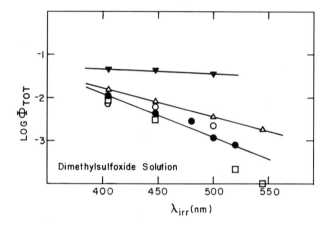

Figure 9. Wavelength dependence of log $\Phi(spec)$ for the photolysis of $Ru(NH_3)_5L^{2+}$ in DMSO solution. Symbols as in Figure 8.

Electron Transfer Photochromism

As mentioned above, the MLCT state for the $Ru(NH_3)_5L^{2+}$ complexes, while unlikely to be substitution active, may be activated toward electron transfer processes with oxidants or reductants in solution. Such bimolecular reactions have been seen for the relatively long-lived MLCT state of $Ru(bipy_3)^{2+}$; thus a comparable pathway might be possible for the relatively substitution-unreactive $Ru(NH_3)_5py\text{-}x^{2+}$ species.

This proposal was tested with the 4-acetylpyridine complex with Cu(II) (0.3M) as the added oxidant in aqueous solution. However, flash photolysis of these solutions (using a conventional xenon flash photolysis apparatus, dead time ~40 μsec) gave no indication of any photoreaction when the wavelengths of irradiation correspond to the MLCT λ_{max}. Neither net nor transient changes in the solution spectrum were noted in contrast to the major decreases in absorbance expected for the shift from left to right of Reaction 9.

$$(NH_3)_5RuN\langle\bigcirc\rangle-C\overset{O^{2+}}{\underset{CH_3}{\diagup}} + Cu^{2+} \underset{}{\overset{h\nu(MLCT)}{\rightleftarrows}} (NH_3)_5RuN\langle\bigcirc\rangle-C\overset{O^{3+}}{\underset{CH_3}{\diagup}} + Cu^{+} \quad (9)$$

Two explanations are consistent: either the outer sphere electron transfer from the MLCT excited state is not occurring significantly, or the electron transfer occurs, but the thermal back reaction to give Ru(II) and Cu(II) is much faster than the response time of the detection apparatus. Several factors support the correctness of the first explanation. Thermal rate constants measured by stopped flow techniques in this laboratory indicate that the back reaction should be rather slow (second-order rate constant < 100 sec^{-1} M^{-1} (14)); thus, significant bleaching of the Ru(II) MLCT band should be easily observable by our flash apparatus. In addition, attempts to measure the lifetime of the weak emission from Ru(NH$_3$)$_5$py-x^{2+} complexes at low temperature gave lifetimes no longer than our excitation/detection dead time ($\sim$20 nsec). Thus, at room temperature, it is probable that the 4-acetylpyridine complex is too short-lived to be quenched significantly by a second-order process.

The pyrazine complex, however, provides a different pathway for electron transfer. The organic ligand has a second coordination site that is a good base for hydrogen ions (19) and for other metal cations (30) (Reaction 10).

$$(NH_3)_5RuN\langle\bigcirc\rangle N^{2+} + M^{n+} \rightleftarrows (NH_3)_5RuN\langle\bigcirc\rangle N\ M^{(n+2)+} \quad (10)$$

$$(M^{n+} = H^+, Cu^{2+}, Zn^{2+}, Ni^{2+}, UO_2^{2+})$$

Under continuous photolysis, the pyrazine complex and its protonated analog are rather photosubstitution unreactive. Also, flash photolysis of these and of the Zn^{2+} adduct (Reaction 10) give no evidence of the formation of transient species with lifetimes longer than 40 μsec. However, when M^{n+} is Cu^{2+} or UO$_2^{2+}$, transient bleaching is observed, and the extent of bleaching is directly proportional to the percent of the Ru(II) complex present as the binuclear adduct. Thus, the bleaching pathway is interpreted as representing the electron transfer process:

$$(NH_3)_5Ru^{II}N\langle\bigcirc\rangle N\ Cu^{II} \overset{h\nu}{\longrightarrow} (NH_3)_5Ru^{III}N\langle\bigcirc\rangle N\ Cu^{I} \quad (11)$$
$$\quad\quad\quad A \quad\quad\quad\quad\quad\quad\quad\quad\quad\quad B$$

The system undergoes relaxation back to starting material; therefore, it is an example of electron transfer photochromism. On the basis of preliminary studies, we concluded (14) that B, which represents a precursor complex of an inner sphere electron transfer reaction, undergoes unimolecular relaxation to starting material. Subsequent studies (17) have shown this conclusion to be incorrect; B apparently undergoes dissociation (Reaction 12) although the eventual return to starting materials occurs via reassociation and inner sphere electron transfer (Reaction 13).

$$(NH_3)_5Ru^{III}N\bigcirc NCu^I \rightleftharpoons (NH_3)_5Ru^{III}N\bigcirc N + Cu^I \quad (12)$$

$$(NH_3)_5Ru^{III}N\bigcirc NCu^I \longrightarrow (NH_3)_5Ru^{II}N\bigcirc NCu^{II} \quad (13)$$

The photolytic-induced electron transfer demonstrates that the MLCT state has a chemistry that is characteristic of its electronic configuration. The quantum yield of this process has a minimum value of 0.12 mol/einstein, yet there is no apparent electron exchange except that which can be attributed to the dinuclear complex A. The present information, however, does not differentiate between mechanisms where electron transfer involves a single concerted step on absorption of light or where electron transfer is first to ligand-localized π^* orbitals, then in a second step to the copper orbitals.

Summary

This presentation has reviewed the aspects of the photochemistry of pentaammineruthenium(II) complexes directly concerned with charge transfer states. In the UV range, photolysis generally leads to the formation of Ru(III) products with concomitant production of H_2 in aqueous solution. This pathway is attributed to charge transfer to solvent processes and the reactive species produced have been shown to be affected by trapping and scavenging agents. Photoaquation from ligand field excited states is an important underlying process at most wavelengths, although the severe drop in Φ_{aq} at λ_{irr} where photooxidation is important suggests that deactivation from the CTTS state occurs via pathways independent of the lower energy LF state.

Complexes having π-unsaturated ligands such as a substituted pyridine display metal-to-ligand charge transfer absorptions in the visible spectral region. The excited state produced by irradiation with these is apparently unreactive toward substitution; it can, however, undergo efficient internal conversion to give aquation active ligand field states. The energy of these MLCT states can be tuned by objective variation of systemic parameters (10, 11, 15) (e.g., ligand substituents or solvent) in a manner which leads to major modifications of the photochemical properties. Among those modifications is the observation that the proximity of an oxidizing metal atom can result in photostimulated electron transfer as the principal reaction mode.

Although the observations reported here are for one series of closely related complexes, clearly the view that subtle ligand modifications may have significant effects on photoactivity has broader implications. In certain cases, such modifications may merely reflect changes in the various deactivation pathways available to reactive excited states; however, the cases where the largest effects may be realized are those where excited states of different orbital parentages have comparable energies within the excited state manifold, especially at the lowest energy range of this manifold.

Acknowledgment

The experimental work reviewed in this presentation was largely carried out by my colleagues: Ray E. Hintze, Tadashi Matsubara, Vincent A. Durante, and George M. Malouf. This work was supported by the National Science Foundation.

Literature Cited

1. Ford, P. C., Petersen, J. D., Hintze, R. E., *Coord. Chem. Rev.* (1974) **14**, 67.
2. Whitten, D. G., Delaive, P. J., Lee, J. T., Abruna, H., Sprintschnik, H. W., Meyer, T. J., ADV. CHEM. SER. (1978) **168**, 28.
3. Watts, R. J., Harrington, J. S., Van Houten, J., ADV. CHEM. SER. (1978) **168**, 57.
4. Sutin, N., Creutz, C., ADV. CHEM. SER. (1978) **168**, 1.
5. Demas, J. N., Adamson, A. W., *J. Am. Chem. Soc.* (1971) **93**, 1800.
6. Gafney, H., Adamson, A. W., *J. Am. Chem. Soc.* (1972) **94**, 8238.
7. Ford, P. C., Stuermer, D. H., McDonald, D. P., *J. Am. Chem. Soc.* (1969) **91**, 6209.
8. Ford, P. C., Chaisson, D. A., Stuermer, D. H., *Chem. Commun.* (1971) 530.
9. Chaisson, D. A., Hintze, R. E., Stuermer, D. H., Petersen, J. D., McDonald, D. P., Ford, P. C., *J. Am. Chem. Soc.* (1972) **94**, 6665.
10. Malouf, G., Ford, P. C., *J. Am. Chem. Soc.* (1974) **96**, 601.
11. Ibid. (1977) **99**, 7213.
12. Hintze, R. E., Ford, P. C., *Inorg. Chem.* (1975) **14**, 1211.
13. Hintze, R. E., Ford, P. C., *J. Am. Chem. Soc.* (1975) **97**, 2664.
14. Durante, V. A., Ford, P. C., *J. Am. Chem. Soc.* (1975) **97**, 6898.
15. Ford, P. C., Malouf, G., Petersen, J. D., Durante, V. A., ADV. CHEM. SER. (1976) **150**, 187.
16. Malouf, G., Ph.D. dissertation, University of California, Santa Barbara, 1977.
17. Durante, V. A., Ph.D. dissertation, University of California, Santa Barbara, 1977.
18. Matsubara, T., Ph.D. dissertation, University of California, Santa Barbara, 1977.
19. Ford, P. C., *Coord. Chem. Rev.* (1970) **5**, 75.
20. Matsubara, T., private communication.

21. Zarnegar, P. P., Bock, C. R., Whitten, D. G., *J. Am. Chem. Soc.* (1973) **95**, 4367.
22. Wrighton, M. S., Morse, D. L., Pdungsap, L., *J. Am. Chem. Soc.* (1975) **97**, 2073.
23. Creutz, C., Sutin, N., *Inorg. Chem.* (1976) **15**, 496.
24. Zanella, A., Ford, K. H., Ford, P. C., *Inorg. Chem.*, in press.
25. Guttel, G., Shiron, M., *J. Photochem.* (1972) **1**, 197.
26. Fox, M., "Concepts of Inorganic Photochemistry," A. W. Adamson, P. D. Fleishauer, Eds., Chapter 8, Wiley–Intersciences, New York, 1975.
27. Hart, E. J., Anbar, M., "The Hydrated Electron," Wiley–Interscience, New York, 1970.
28. Fleischauer, P. D., Adamson, A. W., Satori, G., *Prog. Inorg. Chem.* (1972) **17**, 1.
29. Malouf, G., Ford, P. C., Watts, R. J., unpublished data.
30. Pereira, M. S., Malin, J. M., *Inorg. Chem.* (1974) **13**, 386.
31. Watts, R. J., Crosby, G. A., *J. Am. Chem. Soc.* (1972) **94**, 2606.
32. Lever, A. B. P., "Inorganic Electronic Spectroscopy," Elsevier, New York, 1968.
33. Peterson, J. D., Watts, R. J., Ford, P. C., *J. Am. Chem. Soc.* (1976) **98**, 3188.
34. Nordmeyer, F., Taube, H., *J. Am. Chem. Soc.* (1968) **90**, 1162.

RECEIVED September 20, 1977.

6

Solution Medium Effects on the Photophysics and Photochemistry of Polypyridyl Complexes of Chromium(III)

MARIAN S. HENRY and MORTON Z. HOFFMAN

Department of Chemistry, Boston University, Boston, MA 02215

Photophysical and photochemical studies have been performed on $Cr(bpy)_3^{3+}$ and $Cr(phen)_3^{3+}$ in aqueous solution in the presence of inorganic salts, in neat nonaqueous solvents, and in mixed solvent systems. By use of absorption and emission spectra and flash photolysis, the effects of solution medium (solvent, salts) on excited states and reaction intermediates have been determined. High salt concentrations decrease the lifetime of the ^2E state by decreasing the rate constants for both reactive and nonradiative paths. Non-aqueous solvents do not affect the decay of ^2E but decrease the intersystem crossing efficiency from $^4T_2$ to ^2E. In addition, $^4T_2$ is highly susceptible to solvent attack to form the same seven-coordinate intermediate that arises from reactive decay of ^2E. The chemistry of this intermediate is strongly dependent on the solution medium.

Interest in the photochemistry and photophysics of polypyridyl complexes of Cr(III) (of the form $Cr(NN)_3^{3+}$) has been intensified by the discovery that their lowest excited states have rather long lifetimes and are highly reactive towards redox quenchers. $Cr(bpy)_3^{3+}$ (bpy = 2,2'-bipyridine) and $Cr(phen)_3^{3+}$ (phen = 1,10-phenanthroline) exhibit relatively strong luminescence spectra in aqueous solution at room temperature, each consisting of bands at 695 and 727 nm, and 700 and 727 nm, respectively (1, 2). These emission bands correspond to radiative transitions between the thermally equilibrated (2, 3) $^2T_1$ and ^2E excited states and the $^4A_2$ ground state. The quantum yields of the $^2E \to {}^4A_2$ phosphorescence are $< 10^{-3}$ (3, 4). The lifetimes of these doublet states in nitrogen-purged aqueous solution at 22°C are 0.063 msec for $Cr(bpy)_3^{3+}$ (5) and 0.36 msec for $Cr(phen)_3^{3+}$ (6). The ^2E state (the thermally equilibrated ^2E and $^2T_1$ states will be designated ^2E for simplicity) of $Cr(bpy)_3^{3+}$ is quenched by O_2 ($k = 1.7 \times 10^9$ M^{-1} sec^{-1}) (3), I$^-$ ($k = 1.2 \times 10^9$ M^{-1} sec^{-1}) (3), and Fe_{aq}^{2+} ($k = 4.1 \times 10^7$ M^{-1} sec^{-1}) (7); similar behavior is exhibited by $Cr(phen)_3^{3+}$ (4).

The energy level diagram for $Cr(bpy)_3^{3+}$, shown in Figure 1, has been constructed from the spectral analysis by König and Herzog (8);

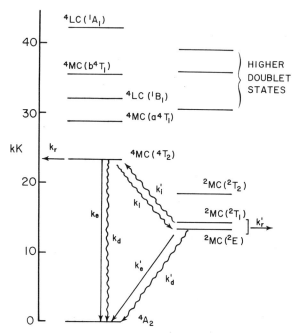

Figure 1. Energy level diagram of $Cr(bpy)_3^{3+}$; MC = metal-centered states, LC = ligand-centered states. k_e, k_d, k_r, and k_i are the rate constants of emission, internal conversion, reaction, and intersystem crossing for the $^4T_2$ state; the primed quantities are the corresponding rate constants for the thermally equilibrated ^2E and $^2T_1$ excited states.

the location of the higher doublet states has been obtained from flash photolysis data (3). However, the lack of a reliable fluorescence emission precludes an exact evaluation of the zero vibrational level of the $^4T_2$ state. With only minor variations in the energies of the states, the diagram for $Cr(phen)_3^{3+}$ is identical. The lack of a Stokes shift between the $^4A_2 \longleftrightarrow {}^2E$ absorption and emission maxima indicates that these states, in both complexes, have virtually the same geometry; they also have the same electronic configuration (t_{2g}^3). One striking difference between the complexes is the apparent efficiency of $^4T_2 \to {}^2E$ intersystem crossing: 0.94 for $Cr(bpy)_3^{3+}$ and 0.21 for $Cr(phen)_3^{3+}$ (9). Excitation from the ground state to the spin-allowed metal- or ligand-centered excited states rapidly leads to population of the $^4T_2$ state, followed by intersystem crossing to form the ^2E state, presumably in less than 1 nsec, as has been found for other Cr(III) complexes (10, 11).

For both complexes, continuous photolysis of neutral and basic aqueous solutions leads to the substitution of two molecules of H_2O or OH^- for a bidentate polypyridyl ligand; both complexes are virtually

$$Cr(NN)_3^{3+} \xrightarrow[H_2O, OH^-]{h\nu} Cr(NN)_2(OH)_2^+ + NN$$

photoinert ($\Phi < 10^{-3}$) in acidic media (3, 4). The limiting value of Φ in air-saturated alkaline solution (Φ_{air}) at 11°C is 0.11 for $Cr(bpy)_3^{3+}$ (3) and ~ 0.01 $Cr(phen)_3^{3+}$ (4). In both cases, the reactive state leading to the photochemical release of NN is believed to be the ^2E state; quenching of the ^2E luminescence results in the quenching of virtually all ($> 97\%$) of the photoreaction (3). Furthermore, both the photochemical (3, 4) and the thermal (12, 13) aquation reactions are believed to involve a common ground state intermediate with acid–base properties; a seven-coordinate $Cr(NN)_3(H_2O)^{3+}$ species generated via an association mechanism with ground or excited (^2E)$Cr(NN)_3^{3+}$ has been postulated. The reformation of $Cr(bpy)_3^{3+}$ after the decay of the ^2E state takes place via acid-dependent and acid-independent paths and accounts for the fact that the complexes are thermally as well as photochemically inert in acidic media. Deprotonation of $Cr(NN)_3(H_2O)^{3+}$ is viewed as leading directly to ring opening and, ultimately, to loss of the polypyridyl ligand with no reformation of the substrate.

The proposed mechanism is summarized in the following equations:

$$(^4A_2)Cr(NN)_3^{3+} + h\nu \to \text{spin-allowed states} \to (^4T_2)Cr(NN)_3^{3+}$$

$$(^4T_2)Cr(NN)_3^{3+} \xrightarrow{k_i} (^4T_2)Cr(NN)_3^{3+}$$

$$(^2E)Cr(NN)_3^{3+} \xrightarrow{k_e'} (^4A_2)Cr(NN)_3^{3+} + h\nu$$

$$\xrightarrow{k_d'} (^4A_2)Cr(NN)_3^{3+} \text{ (nonradiative)}$$

$$\xrightarrow{k_r'} Cr(NN)_3(H_2O)^{3+}$$

$$\mathrm{Cr(NN)_3(H_2O)^{3+}} \xrightleftharpoons{K_a} \mathrm{Cr(NN)_3(OH)^{2+}} + \mathrm{H^+}$$

$$\mathrm{Cr(NN)_3(H_2O)^{3+}} \xrightleftharpoons[k_{th}]{k_{-th},\, \mathrm{H^+}} (^4A_2)\mathrm{Cr(NN)_3^{3+}} + \mathrm{H_2O}$$

$$\mathrm{Cr(NN)_3(OH)^{2+}} \xrightarrow{k_s} \mathrm{Cr(NN_2)(NN\text{-})(OH)^{2+}}$$

$$\xrightarrow{\mathrm{OH^-}} \mathrm{Cr(NN)_2(OH)_2^+} + \mathrm{NN}$$

There have been scattered reports of the effect of solution media on the lifetime and intensity of luminescence and on the quantum yield of the photoreaction of $\mathrm{Cr(bpy)_3^{3+}}$ and $\mathrm{Cr(phen)_3^{3+}}$. Kane-Maquire et al. (2) observed that the addition of DMSO to aqueous solutions of the complexes causes a decrease in the emission intensity that was attributed to suppression of the $^4T_2 \rightarrow {}^2E$ intersystem crossing efficiency. At the same time, they observed that the position of the emission bands and the phosphorescence activation energy remain unchanged (1, 2). Porter reported (14) that the phosphorescence of $\mathrm{Cr(bpy)_3^{3+}}$ is much weaker in neat DMF than in water. We have reported (3) that high concentrations of $\mathrm{ClO_4^-}$ ($> 1M$) cause a significant increase in the emission intensity and lifetime of the ^2E state of $\mathrm{Cr(bpy)_3^{3+}}$ and a decrease in the photochemical quantum yield. Perchlorate ions decrease both the nonradiative (k_d') and reactive (k_r') rate constants of the ^2E states, in concentrated (11.7M) $\mathrm{HClO_4}$, the lifetime in nitrogen-purged solution equals 0.53 msec and 0.67 msec for $\mathrm{Cr(bpy)_3^{3+}}$ and $\mathrm{Cr(phen)_3^{3+}}$, respectively (6).

In this paper, we examine in detail the influence of the solution medium (solvent, salts) on the photophysics and photochemistry of $\mathrm{Cr(bpy)_3^{3+}}$ and $\mathrm{Cr(phen)_3^{3+}}$.

Experimental

Chemicals. Methanol and dimethylformamide (spectroquality) were obtained from Matheson Coleman and Bell and were used without further treatment. Acetonitrile (spectroquality) was obtained from Burdick and Jackson; a fluorescent impurity was removed by two successive distillations with the middle fraction ($\sim$ 70 vol %) reserved each time. Deuterium oxide (99.8%) was obtained from Stohler Isotope Chemicals. $\mathrm{Cr(bipy)_3(ClO_4)_3 \cdot 1/2 H_2O}$ was prepared by the method of Baker and Mehta (15). $\mathrm{Cr(phen)_3(ClO_4)_3 \cdot 2H_2O}$ was the gift of H. R. Hunt, Jr. (16). Other chemicals were reagent grade.

Absorption Spectra. Measurements were made using a Cary 118 recording spectrophotometer. Extinction coefficients ($\pm 3\%$) were measured for both complexes by accurately weighing out a quantity of the salt and dissolving it in a known volume of the appropriate solvent. Measurements were made on solutions of the complexes in neat solvents, 0.10 mol fraction CH_3CN in H_2O, CH_3OH in H_2O, DMF in H_2O, and 0.20 mol fraction DMF in H_2O.

Luminescence. Measurements were performed on air-saturated solutions, unless otherwise noted, using a Perkin–Elmer MPF-2A spectrofluorometer equipped with an R446 or R136 photomultiplier and high sensitivity accessory.

Flash Photolysis. Samples (22-cm pathlength) were purged with high purity nitrogen and replaced after each flash. Flash excitation ($1/e$ time $\sim 30\mu sec$) was filtered through 1 cm of acetone or borosilicate glass. The analyzing beam was passed through a monochrometer ($400 <$ λ, nm < 700) and/or borosilicate glass (< 320 mn). The flash photolysis apparatus has been described previously (17). Laser flash photolysis experiments (1-cm pathlength) were performed in the laboratory of H. Linschitz at Brandeis University. Solutions were air saturated and the excitation wavelength was 347 nm.

Continuous Photolysis. Experiments were performed using a Bausch and Lomb high intensity grating monochromator and a super pressure mercury light source, exciting at 313 nm with a 15-nm bandpass. Quantum yield measurements were made with complete light absorption for $< 10\%$ destruction of $Cr(bpy)_3^{3+}$ and were calculated relative to the reported value for $Cr(bpy)_3^{3+}$ in air-saturated aqueous solution at pH 9.7 and 11°C (3). Solutions were air saturated and stirred during photolysis. The concentration of Cr(III) was obtained as a function of photolysis time by measurement of absorbance of the lowest energy quartet band (~ 450 nm); plots of [Cr(III)] vs. photolysis time were strictly linear. Under these experimental conditions, there was no interference because of product absorption.

Procedures. Mixed solvent systems were prepared by volume assuming no volume change on mixing; measurement of the maximum volume change showed that this approximation introduced a negligible ($< 3\%$) error in the calculation of the mole fraction of the nonaqueous component. A Beckman Expandomatic SS2 instrument was used for pH measurements; pH values assigned to mixed solvent systems were determined on the analogous aqueous solution with H_2O substituted for the nonaqueous component. The anion concentration was assumed to be equal to the formal salt (or acid) concentration, neglecting formation of undissociated ion pairs. Temperature was ambient, $\sim 22°C$.

Results

Ground State Absorption Spectroscopy. The ground state absorption spectrum of $Cr(bpy)_3^{3+}$ in various neat solvents is shown in Figure 2; the results in H_2O are in good agreement with those of König and Herzog (8). The extinction coefficient is solvent dependent, especially in the 250–290 nm region. This is in contrast to the behavior of the free ligand; measurements of the absorption spectrum of bpy in aqueous solution (pH 11.7), in neat CH_3CN, CH_3OH, and DMF show that

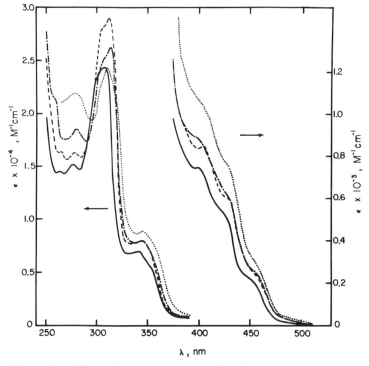

Figure 2. Absorption spectrum of $Cr(bpy)_3^{3+}$ in neat H_2O (———), CH_3CN (– – –), CH_3OH (— — —), and DMF (· · ·); 22°C

variation in ϵ is less than 10%. In aqueous solution, the ground-state absorption spectrum of $Cr(bpy)_3^{3+}$ is dependent upon the concentration of $HClO_4$, as shown in Figure 3.

In dilute aqueous solution, the spectrum of $Cr(phen)_3^{3+}$ is similar [λ_{max}, nm (ϵ, M^{-1} cm^{-1})]: 430(5.9 × 10²), 355(3.9 × 10³), 320(1.2 × 10⁴), 267(6.3 × 10⁴), 225(8.3 × 10⁴), 202(9.8 × 10⁴). These results are in excellent agreement with available literature values (18).

Luminescence Spectroscopy. The phosphorescence spectrum of $Cr(bpy)_3^{3+}$ in nitrogen-purged aqueous solution at 22°C and 77°K is shown in Figure 4. The phosphorescence spectrum of $Cr(phen)_3^{3+}$ is very similar (λ_{max}700, 728 nm). These results are in excellent agreement with published values of λ_{max} for both complexes (2). The excitation spectrum (uncorrected) of the $Cr(bpy)_3^{3+}$ emission (Figure 5) reflects the absorption spectrum. The phosphorescence spectrum at 22°C also is observed from $Cr(bpy)_3^{3+}$ solutions in neat DMF, CH_3OH, and CH_3CN, as well as from the solid salt, with no change ($\pm$ 2nm) in λ_{max}. The wavelength maximum also is unshifted in $HClO_4$ ($\leq 11.7M$). At 77°K in neat H_2O, DMF, CH_2OH, or CH_2CN, the smaller peak at 695 nm disappears, and at least two vibrational bands can be resolved to the red of the main emission band, which remains unshifted; similar results are found for $Cr(phen)_3^{3+}$. The position of the main emission band at 728 nm is unshifted in the presence of up to 0.07 mol fraction DMF (22°C, 77°K) or up to 11.7M $HClO_4$ (22°C, 77°K). In dilute aqueous solution at 22°C with the concentration adjusted to give equal light absorption, the phosphorescence quantum yield of $Cr(phen)_3^{3+}$ is 1.7 times greater than that of $Cr(bpy)_3^{3+}$.

In air-saturated binary solvents (CH_3OH/H_2O, CH_3CN/H_2O, DMF/H_2O) at constant $Cr(bpy)_3^{3+}$ concentration, the phosphorescence intensity decreases with increasing mole fraction of nonaqueous component, χ_s, as shown in Figure 6. The slight increase in the ground state extinction coefficient over this mole fraction range is not sufficient to

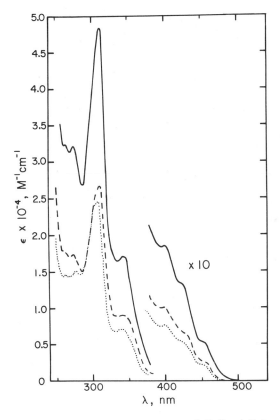

Figure 3. Absorption spectrum of $Cr(bpy)_3^{3+}$ in aqueous $HClO_4$ solution. $[HClO_4] = 1 \times 10^{-4}M$ (· · ·), 0.4M (– – –), 11.7M (———). Right-hand portion of the spectrum has been expanded by a factor of 10.

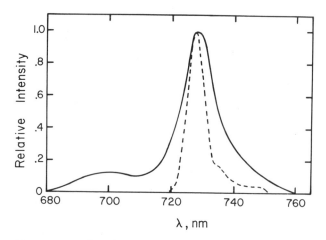

Figure 4. Phosphorescence emission spectrum of Cr-$(bpy)_3^{3+}$ in nitrogen-purged aqueous solution. [Cr-$(bpy)_3^{3+}$] = 1.6 × 10^{-5}M, pH 3.1 in 1 × 10^{-2}M phosphate; λ_{excit} 320 nm; emission band pass 2 nm; 22°C (———), 77°K (- - -).

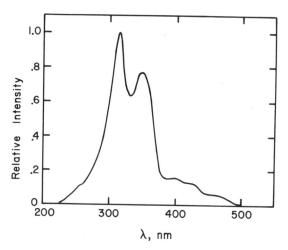

Figure 5. Excitation spectrum (uncorrected) of the phosphorescence of Cr$(bpy)_3^{3+}$ in nitrogen-purged aqueous solution. [Cr$(bpy)_3^{3+}$] = 1.6 × 10^{-5}M, pH 3.1 in 1 × 10^{-2}M phosphate, 22°C.

cause any significant distortion in these data. This decrease in phosphorescence intensity occurs with no change of relative peak heights in the emission or excitation spectra, and is not accompanied by any new fluorescence or phosphorescence (400–800 nm). In DMF/H₂O solutions, the dependence of the phosphorescence intensity of Cr(phen)$_3^{3+}$ on χ_s

is indistinguishable from that of $Cr(bpy)_3^{3+}$. Decreased phosphorescence from $Cr(phen)_3^{3+}$ also is unaccompanied by the appearance of any new emission. In CH_3CN, the phosphorescence intensity of $Cr(bpy)_3^{3+}$ decreases with increasing mole fraction DMF (Figure 7) in a manner quite similar to that found in binary solvents containing water.

Addition of $10^{-3}M$ NaI to a DMF/H_2O solution ($\chi_s = 0.2$, pH 9.8) of $Cr(bpy)_3^{3+}$ completely quenches the phosphorescence. In air-saturated

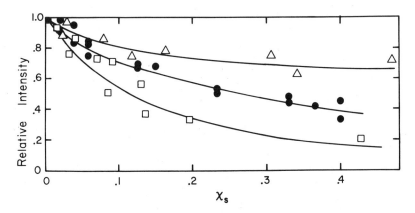

Figure 6. Dependence of phosphorescence emission intensity of Cr-$(bpy)_3^{3+}$ on the mole fraction of nonaqueous component (χ_s) in H_2O: CH_3CN (△), CH_3OH (●), DMF (□); 22°C, air saturated

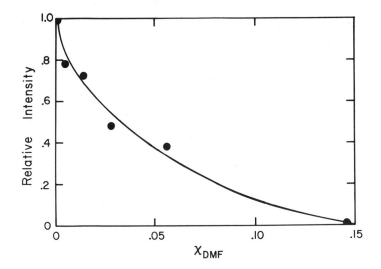

Figure 7. Dependence of phosphorescence emission intensity of $Cr(bpy)_3^{3+}$ on the mole fraction of DMF in CH_3CN; 22°C, air saturated

solution at 22°C, an increase in the concentration of ClO_4^- from $0.40M$ to $11.7M$ causes an increase in the phosphorescence intensity of $Cr(bpy)_3^{3+}$ by a factor of three. The concentration of complex was adjusted to give equal light absorption (integrated area under the absorption spectrum over the excitation bandpass), compensating for the ClO_4^- dependence of ϵ.

Flash Photolysis. Flash excitation of an aqueous solution of $Cr(bpy)_3^{3+}$ produces a transient absorption spectrum (λ_{max}390, 445, 590 nm) that has been identified as the ^2E state of the complex (3). A similar spectrum, slightly blue-shifted, is observed for $Cr(phen)_3^{3+}$ (λ_{max}370, 410, 520 nm). In nitrogen-purged binary solvents (CH_3CN/H_2O, CH_3OH/H_2O, DMF/H_2O) at constant $Cr(bpy)_3^{3+}$ concentration, the decrease in the intensity of the ^2E state absorption signal (measured at the most intense peak, 390 nm) with increasing χ_s parallels the decrease in ^2E state emission intensity shown in Figure 6, indicating that O_2 plays no role in the solvent effect. Similar behavior is observed in C_2H_5OH/H_2O ($\chi_s = 0.067$) and 2-propanol/H_2O ($\chi_s = 0.052$) solutions. Measurements of the ^2E state transient absorption spectrum at values of χ_s, for which the transient absorbance is decreased to one-third its value in aqueous solution, show that the 390-nm band is unshifted in the presence of the nonaqueous component. Measurements of the ground state absorption spectrum show that the decreased ^2E state signal is not caused by decreased absorption of flash excitation ($\lambda > 320$ nm). In 0.2 mol fraction DMF in H_2O, loss of ^2E state absorption is not accom-

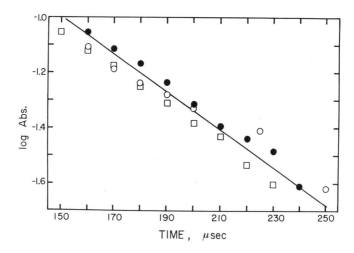

Figure 8. First-order plot of the decay of the ^2E state of $Cr(bpy)_3^{3+}$ in aqueous solution at pH -0.4 (●), 5.5 (□), and 7.9 (○); 22°C, nitrogen-purged solution, λ 390 nm

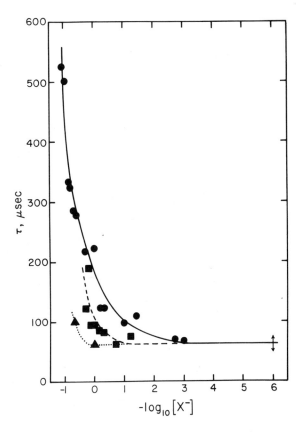

Figure 9. Lifetime of ^2E state of $Cr(bpy)_3^{3+}$ as a function of the concentration of added anions: ClO_4^- (●), HSO_4^- at pH 1 (■), Cl^- (▲); 22°C, nitrogen-purged aqueous solution. Double-headed arrow represents the standard deviation of τ' in dilute solutions.

panied by the appearance (after the 30 μsec flash) of any new transient absorption (360–700 nm). In neat DMF under pulsed laser excitation, the ^2E state absorption is too weak to be detected, and no new transient absorption is observed within 5 μsec after the flash; in aqueous solution the ^2E state absorption is detected readily.

In nitrogen-purged aqueous solution at 22°C, the transient absorption spectrum of the ^2E state of $Cr(bpy)_3^{3+}$ decays via first-order kinetics with $k = 1.6 \times 10^4$ sec^{-1} (Figure 8). Within one standard deviation (0.3 × 10^4 sec^{-1}), the decay rate is unchanged in the following solvent systems: D_2O; ethylene glycol; CH_3CN/H_2O ($0 \leq \chi_s \leq 1$); CH_3OH/H_2O ($0 \leq \chi_s \leq 1$); DMF/H_2O ($0 \leq \chi_s \leq 0.1$); ethanol/H_2O ($\chi_s = 0.067$); 2-propanol/H_2O ($\chi_s = 0.052$).

In nitrogen-purged aqueous solution at 22°C, the ^2E state of Cr(phen)$_3^{3+}$ decays via first-order kinetics with $k = 2.8 \times 10^3$ sec^{-1} (6). Within 10%, this decay rate is unchanged in neat CH$_3$CN.

The lifetime of the ^2E state of Cr(bpy)$_3^{3+}$ is increased in aqueous solution in the presence of high concentrations ($> 1M$) of inorganic salts. The same general effect is found for ClO$_4^-$, HSO$_4^-$, and Cl$^-$ as shown in Figure 9. The effect of ClO$_4^-$ is unchanged if HClO$_4$ is substituted for NaClO$_4$. This effect on the lifetime also is found for 2.4M KNO$_3$ (96 μsec) and 5M LiCl (91 μsec). The transient absorbance is more intense because of the increased lifetime, but the spectral profile is unaffected by the presence of salt.

The presence of ClO$_4^-$ also increases the lifetime of the Cr(phen)$_3^{3+}$ ^2E state. In 11.7M HClO$_4$ (nitrogen-purged, 22°C), the lifetime is 0.67 msec compared with 0.36 msec in dilute aqueous solution.

Continuous Photolysis. Continuous or repeated flash photolysis of Cr(bpy)$_3^{3+}$ in basic aqueous solution produces marked changes in the ground state absorption spectrum, reflecting the photochemical reaction: decreased absorption in the visible region (quartet bands), a blue-shifting of the first strong UV band, and isosbestic points at 306, 270, 262, and

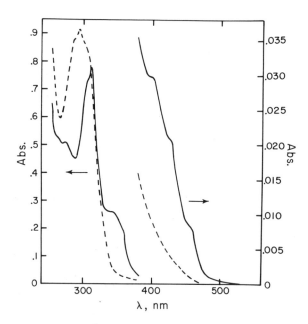

Figure 10. Absorption spectrum (in a 1-cm cell) of $3 \times 10^{-5}M$ Cr(bpy)$_3^{3+}$ before (———) and after (- - -) exhaustive photolysis with a borosilicate glass-filtered xenon arc; 0.2 mol fraction DMF in H_2O, pH 9.8, 22°C, air saturated

Table I. Quantum Yields for Photoreaction of $Cr(bpy)_3^{3+}$ in Various Solvent Systems[a]

Solvent	Conditions	Φ_{air}
H_2O	pH < 4	< 10^{-3}
	pH 9.8	0.16[b]
	pH 9.8, 5M NaCl	0.08
	pH 9.8, 5M NaClO$_4$	0.009
DMF/H_2O ($\chi_s = 0.2$)	pH 1.1	< 10^{-3}
	pH 4.7	0.15
	pH 9.8	0.16
	pH 9.8, $10^{-3}M$ NaI	0.09
	pH 9.8, $10^{-2}M$ NaI	0.07
DMF	neat solvent	0.02
CH_3CN	neat solvent	< 10^{-3}

[a] Air-saturated solutions; 22°C; λ_{excit} 313 nm.
[b] $\Phi_{air} \sim 0.01$ for $Cr(phen)_3^{3+}$ at 11°C under the same conditions (4).

254 nm (3). This reaction is quenched in the presence of acid or I⁻. Φ_{air} for $Cr(bpy)_3^{3+}$ in aqueous (pH 9.7) air-saturated solution at 22°C (λ_{excit} 313 nm) is 0.16, measured relative to the value reported at 11°C (3).

In 0.2 mol fraction DMF in H_2O (pH 9.8), quite similar permanent spectral changes are observed upon continuous or repeated flash photolysis. In Figure 10, the absorption spectrum of a $Cr(bpy)_3^{3+}$ solution is compared with that of the photoinert product solution after exhaustive continuous photolysis. The reaction is partially quenched in the presence of I⁻ and completely quenched in the presence of acid. Continuous or repeated flash photolysis in neat DMF produces spectral changes similar to those in DMF/H_2O. The reaction is not quenched in 0.13M I⁻. Similar spectral changes also are observed for continuous photolysis in neat CH_3CN. In neat CH_3OH, in the absence of oxygen, formation of Cr(II) is observed upon continuous (19) or flash photolysis; in air-saturated solution, no Cr(II) is detected. Because of the formation of Cr(II), the photochemistry of $Cr(bpy)_3^{3+}$ in neat CH_3OH was not investigated further.

Spectral changes described above also are seen upon photolysis of aqueous solutions of $Cr(bpy)_3^{3+}$ containing 5M NaCl or 5M NaClO$_4$. However, Φ_{air} is decreased sharply in these solutions (6). The quantum yield results are summarized in Table I.

Discussion

The basis for the discussion of the effect of the solution medium on the photophysics and photochemistry of $Cr(bpy)_3^{3+}$ and $Cr(phen)_3^{3+}$ is the scheme in Figure 1. The discussion will deal with the rate constants of the various processes as defined in that figure. In addition, $\tau(= 1/(k_e$

$+ k_d + k_r + k_i))$ and $\tau'(= 1/(k_e' + k_d' + k_r' + k_i'))$ represent the lifetimes of the $^4T_2$ and ^2E states, respectively; the efficiency of the jth process from $^4T_2$ and ^2E is designated as $\eta_j(\equiv k_j\tau)$ and $\eta_j'(\equiv k_j'\tau')$, respectively.

Excitation of $Cr(bpy)_3^{3+}$ or $Cr(phen)_3^{3+}$ into spin-allowed metal- or ligand-centered excited states leads to the formation of the $^4T_2$ state. In H_2O, no fluorescence from $^4T_2$ is observed so that $\eta_e = 0$. For $Cr(bpy)_3^{3+}$, $\tau \leq 10^{-9}$ sec, $\eta_r < 0.03$ (3), and $\eta_i = 0.94$ (10); is negligible. Therefore, ^2E is populated rapidly and almost quantitatively upon excitation of that complex in H_2O. For $Cr(phen)_3^{3+}$, $\eta_i = 0.21$ (10) and $\eta_d \sim 0.8$.

For both $Cr(bpy)_3^{3+}$ and $Cr(phen)_3^{3+}$, deactivation of ^2E by radiative decay is unimportant ($\eta_e < 10^{-3}$) compared with nonradiative and reactive modes; also, $\eta_i' \sim 0$ (3). Our study of the photochemical reaction

Table II. Photophysical and Photochemical Constants for ^2E States in Water

	$Cr(bpy)_3^{3+}$	$Cr(phen)_3^{3+}$
τ'_{air} [a]	4.8×10^{-5} sec	7.1×10^{-5} sec
τ' [b]	6.3×10^{-5} sec	3.6×10^{-4} sec
k_r' [c]	3.4×10^3 sec^{-1}	1.4×10^2 sec^{-1}
k_d' [d]	1.3×10^3 sec^{-1}	2.6×10^3 sec^{-1}
η_e'	$< 10^{-3}$	$< 10^{-3}$
η_r' [e]	~ 0.2	~ 0.2
η_d' [f]	~ 0.8	~ 0.8

[a] ^2E lifetime in air-saturated solution at 22°C.
[b] ^2E lifetime in nitrogen-purged solution at 22°C.
[c] $k_r' = \Phi_{air}/\tau'_{air}$; estimated uncertainty $\pm 25\%$.
[d] $k_d' = \tau'^{-1} - k_r'$; estimated uncertainty $\pm 30\%$.
[e] $\eta_r' = \Phi/\eta_i$; $\Phi = k_r'\tau'$.
[f] $\eta_d' = 1 - \eta_r'$.

mechanism of $Cr(bpy)_3^{3+}$ (3) indicates that in alkaline H_2O, formation of the seven-coordinate intermediate via k_r' leads directly and irreversibly to ligand loss and to the $Cr(NN)_2(OH)^{2+}$ product. Therefore, $\eta_r' = \Phi/\eta_i$. In air-saturated solution, the quantum yield of product formation and the efficiency of the reactive pathway leading to the seven-coordinate intermediate are lowered because of the reaction of ^2E with oxygen. However, η_r' and η_d' can be evaluated from Φ_{air}, τ_{air}', and τ'. The photochemical parameters for $Cr(bpy)_3^{3+}$ and $Cr(phen)_3^{3+}$ in H_2O are compared in Table II.

Compared with the lifetimes of ^2E states of other octahedral nitrogen-bonded $Cr(III)$ complexes in aqueous solution ($\tau' < 2 \mu sec$) (20) (the emission lifetime of the ^2E state of $trans$-$Cr(en)_2(NCS)_2^+$ recently has been found to be $\sim 10\mu sec$ (21)), the ^2E states of $Cr(bpy)_3^{3+}$ and $Cr(phen)_3^{3+}$ are remarkably long lived. Inasmuch as $k_d' > (k_e' + k_r' +$

k_i'), the nonradiative return of ^2E to the ground state is the predominant factor determining the value of τ'. There is no isotope effect on τ' for the substitution of D_2O for H_2O as the solvent, indicating the lack of direct vibrational coupling (hydrogen-bonding) between the complex and the bulk solvent. As has been pointed out before (6) and further demonstrated here, τ' (and hence k_d') is not sensitive to changes in solvent polarity (within experimental error), unlike $Cr(CN)_6^{3-}$ (22), which reflects the insulation of the metal-centered ^2E state from electric dipole perturbation by the bulk solvent (23). Thus, the ^2E state is independent of solvent, and the nonradiative transformation of metal-centered electronic energy to ligand-centered vibrational energy depends only on ligand vibrations. In this instance, the ligand vibrational modes are acting as the energy sink as well as providing the oscillating dipole perturbation that permits the energy transfer process to occur.

Both k_r' and k_d' for $Cr(phen)_3^{3+}$ are significantly smaller than the corresponding values for $Cr(bpy)_3^{3+}$. Clearly, substitution of the phen ligands for bpy retards both the reaction of ^2E with H_2O to form the seven-coordinate intermediate and the nonradiative transition to the ground state. The bulkier, more rigid phen ligands should impede the entrance of a water molecule into the pockets between the ligands, which is necessary to effect coordination. Similarly, this increase in rigidity upon substituting phen for bpy is reflected in the low-energy metal–nitrogen vibrational modes which are at a higher frequency for tris(phen) complexes than for tris(bpy) complexes (24). Substitution of phen for bpy very sharply restricts the vibrational modes involving stretching, bending, or rotation about the inter-ring C–C bond. Because the Cr–N bonds and the ligand dipole moment should be very sensitive to these vibrations, the extent of the phen perturbation on the ^2E state is expected to be much smaller relative to bpy, and thus, in $Cr(phen)_3^{3+}$, the rate of energy transfer is decreased. Incorporation of substituents on the bpy, phen, or other polypyridyl ligands is expected to cause systematic changes in k_i' and k_d' so that τ' of these complexes can be engineered and fine tuned (25).

Effect of Anions. The prolongation of τ' in the presence of high concentrations of anions is in direct contrast to the phenomenon of quenching in which the added solute causes the lifetime of an excited state to be decreased. Clearly, the anions used in this study do not quench the ^2E states or, if they do, that effect is overwhelmed by the prolongation effect. From the values of τ', τ_{air}', and Φ_{air} given in "Results," the photophysical and photochemical constants of the complexes in the presence of high concentrations of ClO_4^- and Cl^- have been evaluated (Table III). It is important to note that inasmuch as the energy of the ^2E states is not affected by the presence of the anions, η_i' still is deemed to be negligible. Similarly, although η_e' increases slightly as the anion concentration is increased, as a direct result of the increase in τ', η_e' remains negligible compared with η_r' and η_d'.

It can be seen readily by comparing Tables II and III that the increase in τ' in the presence of anions results from the decrease in both k_r' and k_d'. In the 5M salt solutions, k_r' is more highly affected than is k_d'; in 11.7M HClO$_4$, the major contributor to the prolongation of τ' certainly appears to be k_d'. The magnitude of this effect and the lifetimes of the excited states that result are viewed as rather extraordinary.

There is no doubt that ion pairing between the anions and the cationic complexes is extensive, as evidenced by the dependence of the ground state extinction coefficient of Cr(bpy)$_3^{3+}$ on [ClO$_4^-$]. For the Fe(phen)$_3^{2+}$

Table III. Photophysical and Photochemical Constants for ^2E States in Concentrated ClO$_4^-$ and Cl$^-$ Aqueous Solutions

	$Cr(bpy)_3^{3+}$			$Cr(phen)_3^{3+}$
	5M $NaCl$	5M $NaClO_4$	11.7M $HClO_4$[g]	11.7M $HClO_4$[g]
τ'_{air}[a]	4.5 × 10^{-5} sec	7.1 × 10^{-5} sec	nd	nd
τ'[b]	1.0 × 10^{-4} sec	3.1 × 10^{-4} sec	5.3 × 10^{-4} sec	6.7 × 10^{-4} sec
k_r'[c]	1.7 × 10^3 sec^{-1}	1.2 × 10^2 sec^{-1}	negligible	negligible
k_d'[d]	8 × 10^3 sec^{-1}	3 × 10^3 sec^{-1}	1.9 × 10^3 sec^{-1}	1.5 × 10^3 sec^{-1}
η_r'[e]	~ 0.2	~ 0.04	negligible	negligible
η_d'[f]	~ 0.8	~ 0.96	~ 1	~ 1

[a] ^2E lifetime in air-saturated solution at 22°C.
[b] ^2E lifetime in nitrogen-purged solution at 22°C.
[c] $k_r' = \Phi_{air}\tau'_{air}$; estimated uncertainty ± 25%.
[d] $k_d' = \tau'^{-1} - k_r'$; estimated uncertainty ± 30%.
[e] $\eta_r' = k_r'\tau'$.
[f] $\eta_d' = 1 - \eta_r'$.
[g] The complexes are photoinert in acidic solution because of the H$^+$-promoted conversion of the seven-coordinate intermediate to the substrate; therefore $\Phi_{air} = 0$. In 11.7M HClO$_4$, formation of the seven-coordinate intermediate (k_r') would be very small (see "Discussion").

··· ClO$_4^-$ ion pair, $K_{ip} = 5.73$ (26); for the tripositive cations in this study, K_{ip} should be even larger for electrostatic reasons. Thus, the ^2E state (as well as the ground state) is tightly bound with anions in these concentrated salt solutions. Also, a considerable amount of bulk solvent is bound up in the solvation of the ions (27), causing a lowering of the activity of H$_2$O and altering its structure. Therefore, it is easy to see how the presence of anions would restrict formation of the seven-coordinate intermediate and cause a decrease in the value of k_r'.

The effect of the anions on k_d' requires a decrease in the rate of energy transfer from the ^2E state into the ligand vibrational modes. A possible model allows the anions to enter somewhat into the inter-ligand pockets (*26*), thereby decreasing the vibrational freedom of the ligands and the efficiency of energy transfer. The model predicts that the effect of anions on τ' should be less pronounced for $Cr(phen)_3^{3+}$ than for $Cr(bpy)_3^{3+}$ because the more rigid phen ligands have less vibrational freedom. Note that the maximum effect on τ' (and hence k_d') is almost a factor of 10 for $Cr(bpy)_3^{3+}$ but only a factor of two for $Cr(phen)_3^{3+}$.

Consideration must be given for a moment to the possible effect of extensive ion pairing on the behavior of the $^4T_2$ state. Even if it is assumed that anions also cause a decrease in k_r and k_d, the net result is no change in η_i from its value of ~ 1 in dilute aqueous solution; k_i remains the lifetime-determining (fastest) step.

In summary, the magnitude of the effect on τ' by the specific anions is the result of composite phenomena: ion pair formation, molecular events in the inter-ligand pockets, vibrational freedom of the ligands, and structure and activity of water. Thus, the results shown in Figure 9 are, unfortunately, not amenable to quantitative treatment at this time. Nevertheless, the prolongation of the lifetimes of the ^2E states of polypyridyl complexes of Cr(III) by high concentrations of anions appears to be a general phenomenon (*25*).

Effect of Solvents. In the presence of nonaqueous solvents, the population of the ^2E state of $Cr(bpy)_3^{3+}$ or $Cr(phen)_3^{3+}$, as reflected by their emission and absorption intensities, is strikingly lower than in neat H_2O. At the same time, their energies, lifetimes, and absorption spectral profiles are unaffected by change in solvent. Inasmuch as τ' is unchanged, diffusional quenching of ^2E by the nonaqueous solvent does not occur. Also, τ' is very long compared with the rate of solvent exchange (*28*), so that ^2E encounters many nonaqueous solvent molecules during its lifetime. This implies that the presence of the nonaqueous solvent component in the first solvation sphere of the ^2E complex is not sufficient for reaction; therefore, the nonaqueous solvent does not introduce any static quenching process for deactivation of ^2E.

The higher excited states which produce ^2E are distinguished by quite different distributions of electron density. Therefore, the strength of interaction with the solvent environment should differ among the various ^2E state precursors. For this reason, if the solvent effect is caused by interception of some precursor state, the interception reaction should be reflected in the shape of the phosphorescence excitation spectrum; as χ_s increases, bands arising from the intercepted state should disappear. No such effect is found experimentally, requiring that the $^4T_2$ state is both the exclusive source of the ^2E state and the only precursor state interacting with the solvent. Thus, as suggested by Kane-Maquire et al. (*2*), the lowering of the population of ^2E results from a decrease in η_i, which, in turn, must arise from an increase in η_d and/or η_r because of interaction of $^4T_2$ with the nonaqueous solvent; η_e remains negligible. To compete with intersystem crossing, any interception reaction must have a pseudo first-order rate constant $k \geq 10^{10}$ sec^{-1}.

The value of the phosphorescence intensity of Cr(bpy)$_3^{3+}$, relative to that in H$_2$O where $\eta_i \sim 1$ (*10*), as a function of χ_s (*see* Figure 6), can be taken as a direct measure of η_i. Thus, in 0.2 mol fraction DMF in H$_2$O, the population of ^2E is about one-third as large as that in neat H$_2$O, and $\eta_i \sim 0.3$.

The ^2E molecules formed from $^4T_2$ in mixed solvent engage in the same relaxation processes as in neat H$_2$O. The presence of $10^{-3}M$ I$^-$ in neat H$_2$O quenches virtually all of the phosphorescence and photochemistry arising from the ^2E state of Cr(bpy)$_3^{3+}$ (*3*). In 0.2 mol fraction DMF in H$_2$O, the presence of $10^{-3}M$ I$^-$ completely quenches the phosphorescence from the ^2E state of Cr(bpy)$_3^{3+}$ but only quenches about one-third of the photoreaction in alkaline solution. The presence of $10^{-2}M$ I$^-$ has hardly any further effect on Φ_{air}. The conclusion is drawn that the quenchable part of the photoreaction ($\Phi_{air} \sim 0.05$) arises from ^2E and that the unquenchable part ($\Phi_{air} \sim 0.01$) arises from a short-lived precursor state. Assuming that the rate constant for the quenching of this precursor state by I$^-$ is $\sim 10^{10}$ M^{-1} sec^{-1} in this solvent, the lifetime of the nonquenchable precursor must be ≤ 10 nsec; τ for $^4T_2$ is ≤ 1 nsec. Thus, these quenching experiments support the assignment of the $^4T_2$ state as the precursor state that interacts with the solvent.

Although excited quartet states of Cr(III) are viewed historically as undergoing dissociative reaction (*29*), the fact that the photoreaction is quenched by acid in 0.2 mol fraction DMF in H$_2$O suggests that $^4T_2$ (as well as ^2E) reacts via an associative mechanism in this binary solvent. The seven-coordinate intermediates from $^4T_2$ and ^2E would most likely be identical, indistinguishable, or interconvertable, owing to the range of geometric and electronic configurations possible for such species (*30*). Therefore, the assumption is made that the seven-coordinate precursor to the photoproducts (Cr(NN)$_3$S^{3+}) is the same irrespective of the state of origin ($^4T_2$ or ^2E).

The value of the photochemical quantum yield depends upon η_r, η_r', and the efficiency of conversion of the seven-coordinate intermediate into the ligand-substituted product (η_s). This latter process, involving ring-opening, ligand loss, and solvent substitution, competes with the loss of the seventh S ligand (η_{-th}) to reform the ground state substrate. In binary solvents, consideration must be given to the solvent exchange reaction of the seven-coordinate intermediate, the efficiency of which is η_x.

$$\text{Cr(NN)}_3\text{S}^{3+} + \text{S}' \xrightleftharpoons{k_x} \text{Cr(NN)}_3\text{S}'^{3+} + \text{S}$$

The efficiencies of the transformations of the seven-coordinate intermediate should be strongly dependent on the solution medium and, specifically, on the nature of S in the seventh coordination site. For example, we know that in alkaline water when S = OH$^-$, $\eta_s = 1$ and $\eta_{\text{-th}} = 0$. On the other hand, when S = H$_2$O, $\eta_s < 1$ and $\eta_{\text{-th}} > 0$ and as [H$^+$] is increased, $\eta_s \to 0$ and $\eta_{\text{-th}} \to 1$ (3).

An evaluation of the efficiencies for Cr(NN)$_3$S^{3+} where S = DMF can be made from an examination of the behavior of the system in that neat solvent. The luminescence from ^2E is lower by more than a factor of 100 from that in neat H$_2$O; clearly, $\eta_i \sim 0$. The observed photoreaction ($\Phi_{\text{air}} = 0.02$) is not quenched by I$^-$ and must originate only from $^4T_2$ inasmuch as virtually no ^2E is present. In neat H$_2$O, $\eta_d \sim 0$; in neat DMF, η_d would also be ~ 0. There does not appear to be any reasonable mechanism for the change of bulk solvent composition to increase the rate of energy transfer from the metal-centered $^4T_2$ into the vibrational modes of the ligand. Notwithstanding the distorted geometry of $^4T_2$ relative to ^2E and $^4A_2$, it should be noted that solvent composition has no effect on k_d' for ^2E. Therefore, in neat DMF, $\eta_r \sim 1$ and the sole mode of decay of $^4T_2$ is transformation to Cr(NN)$_3$(DMF)$^{3+}$. In this neat solvent, the observed quantum yield must arise from the values of η_s (~ 0.02) and $\eta_{\text{-th}}$ (~ 0.98). Thus, unlike the situation where S = OH$^-$, the predominant process for Cr(NN)$_3$(DMF)$^{3+}$ is reversion to the substrate through facile loss of the seventh ligand. Similarly, Φ_{air} in neat CH$_3$CN is $< 10^{-3}$ although $\eta_i \sim 0.5$. The photoinertness of Cr(bpy)$_3^{3+}$ in CH$_3$CN can be attributed to the high efficiency of return of Cr(bpy)$_3$(CH$_3$CN)$^{3+}$ to the ground state substrate.

In 0.2 mol fraction DMF in H$_2$O, Cr(bpy)$_3^{3+}$ is almost completely preferentially solvated by DMF (31). The first solvation sphere of the complex consists mainly of DMF and does not match the composition of the bulk solvent. Because the short lifetime of $^4T_2$ precludes extensive solvent exchange, the seven-coordinate intermediate formed from $^4T_2$ should be Cr(bpy)$_3$(DMF)$^{3+}$ initially. The ^2E lifetime is very long compared with the rate of solvent exchange (27). However, the preferential solvation shown by $^4A_2$ also should occur for ^2E so that the seven-coordinate intermediate formed from ^2E should also be Cr(bpy)$_3$(DMF)$^{3+}$. Inasmuch as $\eta_i \sim 0.3$ and $\eta_d \sim 0$, $\eta_r \sim 0.7$. Furthermore, inasmuch as τ' is not solvent dependent, the values of η_r' and η_d' can be taken as those in neat water (~ 0.2 and ~ 0.8, respectively). The quantum yield of formation of Cr(bpy)$_3$(DMF)$^{3+}$ from ^2E then would be $\eta_i\eta_r' \sim 0.06$; the quantum yield of formation of Cr(bpy)$_3$(DMF)$^{3+}$ from $^4T_2$ would be η_r

~ 0.7. The total quantum yield would be ~ 0.76. For this intermediate, $\eta_s \sim 0.02$. These values predict that the quantum yield of the photoreaction should be ~ 0.015 or approximately a factor of 10 lower than is observed. This difference can be reconciled if one assumes that Cr-(bpy)$_3$(DMF)$^{3+}$ undergoes exchange of the seventh ligand with H$_2$O to form Cr(bpy)$_3$(H$_2$O)$^{3+}$. Deprotonation of the seventh ligand in the alkaline aqueous medium yields Cr(bpy)$_3$(OH)$^{2+}$ for which $\eta_s = 1$. Therefore, if $\eta_x \sim 0.2$ for this DMF–H$_2$O exchange reaction, $\eta_{\text{-th}} \sim 0.8$ for conversion of Cr(bpy)$_3$(DMF)$^{3+}$ to the substrate. The net result of this analysis is the rationalization of the quantum yield of the photochemical reaction ($\eta_x(\eta_r + \eta_i \eta_r)$) and its quenchable and nonquenchable components.

Scheme 1

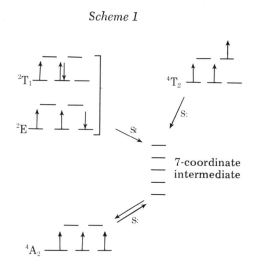

At this time, it appears that the effect of solvent on the $^4T_2$ state is to change the relative efficiencies of the reactive and intersystem crossing paths without affecting the radiative and nonradiative paths. Inasmuch as k_i for Cr(bpy)$_3^{3+}$ should be independent of solvent environment involving, as it does, nonradiative conversion between two metal-centered states, the fundamental solvent effect on $^4T_2$ is to increase k_r. In contrast, the solvent environment does not seem to have an effect on k_r'. The specific values of the quantum yield of the photoreaction appear to be determined by the thermal chemistry of the seven-coordinate intermediate formed from both $^4T_2$ and ^2E.

The interaction of S with $^4T_2$ to form the seven-coordinate intermediate is easy to rationalize on the basis of the electronic configuration of the state. The $^4T_2$ state is distorted relative to the $^4A_2$ ground state because of its $t_{2g}^2 e_g^1$ configuration (Scheme 1). Even in the ground state, the orientation of the three planar polypyridyl ligands defines interligand pockets large enough to accommodate an average of 2.5 H_2O molecules (25); in the $^4T_2$ state, where the metal–ligand bond distances are longer than in the ground state, the interligand pockets should be even larger. The S molecules are characterized by lone pairs of electrons and can act as Lewis bases; the vacant t_{2g} orbital, pointing out into the interligand pockets, can act as the electron acceptor. Coordination of S yields a ground state seven-coordinate intermediate which can have a wide range of geometric orientations. Depending upon the geometry, the five d orbitals become mixed and reordered in energy (30). Loss of S and reordering of the orbital energies yields the six-coordinate quartet ground state substrate in competition with the step that establishes the six-coordinate quartet ring-opened intermediate.

Examination of the orbital configuration scheme reveals the reasons for the relative slowness of the reaction of ^2E with S to form the seven-coordinate intermediate. The ^2E state is geometrically identical to the ground state and not distorted as is $^4T_2$; as well, ^2E does not have a vacant t_{2g} orbital to accommodate S. It is entirely possible that the reactive pathway from ^2E actually occurs through the thermally equilibrated $^2T_1$ state which has the requisite configuration. The barriers to this reaction are reflected in the low values of k_r' relative to k_r. The electron configuration of ^2E is more important in determining the value of k_r' than is the nature of S. Because k_r' is much smaller than k_d' (which is independent of solvent) and probably is only weakly solvent dependent, τ' is experimentally independent of solvent. Similarly, the thermal reaction of S with the ground state would be very slow because of the large configurational and energetic changes required; $k_{th} = 3.3 \times 10^{-6}$ sec^{-1} for Cr-(bpy)$_3{}^{3+}$ in H_2O at 25°C with $\Delta H^\ddagger = 22.3$ kcal mol^{-1} and $\Delta S^\ddagger = -8.8$ eu (12).

It remains only to account for the ordering of the solvents with regard to the lowering of η_i: $H_2O < CH_3CN < CH_3OH < DMF$. Data taken in DMF/CH_3CN solutions confirm this ordering and demonstrate that the effect is not specific to water. The ordering does not correlate with bulk solvent dielectric constant or refractive index or the gas-phase dipole moment. Perhaps a better measure would be Gutmann's donor number (32), which reflects the ability of S to act as a Lewis base. However, the available values refer to dilute solution and apparently are influenced strongly by solvent structure. For example, DN = 18.0 for dilute H_2O dissolved in 1,2-dichloroethane but in bulk solvent, DN ~ 33. More importantly, comparison of donor numbers is invalid for cases in which steric effects are important.

There is no question that the solvent interacts strongly with the ground state complexes. As Figure 2 shows, the absorption spectrum of

$Cr(bpy)_3^{3+}$ is affected by the nature of the neat solvent. While the λ_{max} values for the various absorption bands are not particularly sensitive to solvent, indicating that excited state energies are relatively unchanged, the ϵ values are strongly solvent dependent. In the 250–300 nm region, that corresponds to a "window" in the spectrum, ϵ_{DMF} is about 50% higher than ϵ_{H_2O}. The interaction of $HClO_4$ with the ground state complex, involving ion pair formation (see Figure 3), also causes an increase in ϵ. Of particular interest is that the ϵ values in the 250–300 nm region are in the order $H_2O < CH_3CN < CH_3OH < DMF$, the same order as the effect of solvent on η_i. These results suggest that the functional dependence of η_i on χ_s is regulated largely by the magnitude of preferential solvation of the complex in the binary solvent systems.

It is clear that the effect of solvents on the photophysics and photochemistry of polypyridyl complexes of Cr(III) is very complicated and depends upon preferential solvation of the ground state, interaction of solvents with the $^4T_2$ and $^2E/^2T_1$ states, and the behavior of the seven-coordinate intermediate. In principle, by knowing the quantitative nature of these interactions and the parameters upon which they depend, excited state lifetimes, quantum yields, and photochemical synthetic routes could be engineered.

Conclusions

The photophysical and photochemical mechanism for $Cr(NN)_3^{3+}$ complexes in solvents or anionic solutes (S) is summarized in Scheme 2.

Scheme 2

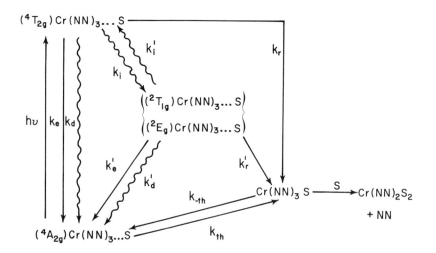

Excitation of the solvated or ion-paired ground state into spin-allowed metal- or ligand-centered excited states leads to the formation of the $^4T_2$ state with unitary efficiency. $^4T_2$ is distorted relative to the ground state and remains solvated or ion paired. Deactivation to the lowest doublet states competes with the reactive mode. Nonaqueous solvents increase k_r; in H_2O the reactive path is negligible compared with intersystem crossing. The ^2E state is not distorted relative to the ground state but remains solvated or ion paired. Anions decrease k_r' and k_d' and thereby prolong the lifetime of ^2E; solvents do not affect the lifetime of ^2E. The reactive pathway from ^2E (which may involve $^2T_1$) and from $^4T_2$ leads to a seven-coordinate intermediate that can revert to the ground state substrate or engage in solvent-coupled processes leading to the final photoproducts. The chemistry of the seven-coordinate intermediate is very dependent upon the nature of the solution medium.

Acknowledgment

Financial support from the North Atlantic Treaty Organization (Grant No. 658) and the National Science Foundation (Grant No. CHE 76-21050) is gratefully appreciated. The authors also acknowledge their many discussions of this work with V. Balzani, N. Serpone, and V. W. Cope.

Literature Cited

1. Kane-Maquire, N. A. P., Langford, C. H., *J. Chem. Soc., Chem. Commun.* (1971) 895.
2. Kane-Maquire, N. A. P., Conway, J., Langford, C. H., *J. Chem. Soc., Chem. Commun.* (1974) 801.
3. Maestri, M., Bolletta, F., Moggi, L., Balzani, V., Henry, M. S., Hoffman, M. Z., *J. Am. Chem. Soc.* (1978) **100**.
4. Maestri, M., personal communication of unpublished data.
5. Maestri, M., Bolletta, F., Moggi, L., Balzani, V., Henry, M. S., Hoffman, M. Z., *J. Chem. Soc., Chem. Commun.* (1977) 491.
6. Henry, M. S., *J. Am. Chem. Soc.* (1977) **99**, 6138.
7. Ballardini, R., Varani, G., Scandola, F., Balzani, V., *J. Am. Chem. Soc.* (1976) **98**, 7432.
8. König, E., Herzog, S., *J. Inorg. Nucl. Chem.* (1970) **32**, 585.
9. Bolletta, F., Maestri, M., Balzani, V., *J. Phys. Chem.* (1976) **80**, 2499.
10. Kirk, A. D., Hoggard, P. H., Porter, G. B., Rockley, M. C., Windsor, M. W., *Chem. Phys. Lett.* (1976) **37**, 193.
11. Castelli, F., Forster, L. S., *Chem. Phys. Lett.* (1975) **30**, 465.
12. Maestri, M., Bolletta, F., Serpone, N., Moggi, L., Balzani, V., *Inorg. Chem.* (1976) **15**, 2048.
13. Serpone, N., Jamieson, M. A., Maestri, M., unpublished data.
14. Porter, G. B., *Microsymposium Photochem. Photophys. Coord. Compounds, Italy, July, 1976.*
15. Baker, B. R., Mehta, B. D., *Inorg. Chem.* (1965) **4**, 848.

16. Lee, C. S., Gorton, E. M., Neumann, H. M., Hunt, H. R., Jr., *Inorg. Chem.* (1966) **5**, 1397.
17. Vaudo, A. F., Kantrowitz, E. R., Hoffman, M. Z., Papaconstantinou, E., Endicott, J. F., *J. Am. Chem. Soc.* (1972) **94**, 6655.
18. Ferguson, J., Hawkins, C. J., Kane-Maquire, N. A. P., Lip, H., *Inorg. Chem.* (1969) **8**, 777.
19. Ballardini, R., Scandola, F., personal communication of unpublished data.
20. Adamson, A. W., Gutierrez, A. R., Wright, R. E., Walters, R. T., *Informal Conf. Photochem., 12th, National Bureau of Standards, Washington, D.C., June, 1976*, paper G1.
21. Sandrini, D., Gandolfi, M. T., Moggi, L., Balzani, V., *Am. Chem. Soc.* (1978) **100**, 1463.
22. Dannöhl-Fickler, R., Kelm, H., Wasgestian, F., *J. Lumin.* (1975) **10**, 103.
23. Merkel, P. B., Kearns, D. R., *J. Am. Chem. Soc.* (1972) **94**, 7244.
24. Ferraro, J. R., "Low Frequency Vibrations of Inorganic and Coordination Compounds," pp. 198–202, Plenum, New York, 1971.
25. Serpone, N., Jamieson, M. A., Henry, M. S., Hoffman, M. Z., "Abstracts of Papers," *National Meeting, Am. Chem. Soc., 175th, March 1978*.
26. Van Meter, F. M., Neumann, H. M., *J. Am. Chem. Soc.* (1976) **98**, 1382.
27. Janz, G. J., Oliver, B. G., Lakshminarayanan, G. R., Mayer, G. E., *J. Phys. Chem.* (1970) **74**, 1285.
28. Behrendt, S., Langford, C. H., Frankel, L. S., *J. Am. Chem. Soc.* (1969) **91**, 2236.
29. Zinato, E., "Concepts of Inorganic Photochemistry," A. W. Adamson, P. D. Fleischauer, Eds., p. 143, John Wiley, New York, 1975.
30. Hoffmann, R., Beier, B. F., Muetterties, E. L., Rossi, A. R., *Inorg. Chem.* (1977) **16**, 511.
31. Langford, C. H., personal communication of unpublished data.
32. Gutmann, V., Schmid, R., *Coord. Chem. Rev.* (1974) **12**, 263.

RECEIVED September 20, 1977.

7

Photochemical Processes in Cyclopentadienylmetal Carbonyl Complexes

DONNA G. ALWAY and KENNETH W. BARNETT

Department of Chemistry, University of Missouri–St. Louis, St. Louis, MO 63121

Photolysis of η^5-$C_5H_5Fe(CO)_2X$ ($X = Cl$, Br, or I) at 366 or 436 nm results in expulsion of a carbon monoxide ligand as the only detectable photochemical process. Irradiation of ^{13}CO-saturated solutions forms η^5-$C_5H_5Fe(^{12}CO)(^{13}CO)X$. In the presence of triphenylphosphine, neutral complexes η^5-$C_5H_5Fe(CO)(PPh_3)X$ ($X = Br$ or I) are formed. Quantum yields for the latter processes are essentially invariant with respect to irradiation wavelength but increase with increasing triphenylphosphine concentration, reaching limiting values of 0.96 and 0.95 for the bromide and iodide, respectively. Irradiation of η^5-$C_5H_5Mo(CO)_3X$ ($X = Cl$, Br, or I) and PPh_3 affords cis-η^5-$C_5H_5Mo(CO)_2(PPh_3)X$, suggesting that the CO ligands cis to halogen are preferentially labilized. Cis- and trans-η^5-$C_5H_5Mo(CO)_2(PPh_3)X$ undergo geometric isomerization and competitive loss of CO and PPh_3.

The utility of visible or UV irradiation as a synthetic tool in organometallic chemistry has been recognized for many years (1, 2, 3). Only recently, however, have investigations of the detailed nature of the processes involved in photochemical activation been reported. For binary metal carbonyls, dissociation of carbon monoxide (Reaction 1) occurs

$$M(CO)_n \xrightarrow{h\nu} M(CO)_{n-1} + CO \qquad (1)$$

with high efficiency (1, 2, 3). The coordinatively unsaturated products, $Fe(CO)_4$ (4) and $M(CO)_5$ (5, 6, 7) [$M = Cr$, Mo, or W], have been observed spectroscopically following photolysis of $Fe(CO)_5$ and the Group VI hexacarbonyls, respectively. Monosubstituted derivatives may

$$M(CO)_nL \xrightarrow{h\nu} M(CO)_{n-1} + CO \qquad (2)$$

$$M(CO)_nL \xrightarrow{h\nu} M(CO)_n + L \qquad (3)$$

dissociate either carbon monoxide or the donor ligand L, depending upon the nature of the latter (Reactions 2 and 3). The factors determining which process will occur are well defined only for a series of $M(CO)_5L$ complexes [M = Mo, W] (1, 8, 9).

Complexes containing metal–metal bonds appear to undergo two competing processes as the result of UV or visible radiation (10, 11, 12, 13, 14). $Mn_2(CO)_{10}$ and $Re_2(CO)_{10}$ undergo homolytic metal–metal bond cleavage to yield the radical species $M(CO)_5$, which are extraordinarily substitution labile (14). The corresponding processes also are

$$M_2(CO)_{10} \xrightarrow{h\nu} 2M(CO)_5 \qquad (4)$$

$$[CpM(CO)_3]_2 \xrightarrow{h\nu} 2CpM(CO)_3 \qquad (5)$$

$$[CpM(CO)_3]_2 \xrightarrow{h\nu} Cp_2M_2(CO)_5 + CO \qquad (6)$$

observed for $[CpMo(CO)_3]_2$ and $[CpW(CO)_3]_2$ (10) $[Cp = \eta^5\text{-}C_5H_5]$, but flash photolysis of the former suggests that CO dissociation to give $Cp_2Mo_2(CO)_5$ (11) also occurs in this case. Recent studies of $CpM(CO)_3$ (15) and $M(CO)_5X$ (16) [M = Mn or Re, X = halide] indicate that CO expulsion is the dominant photochemical process for these complexes.

Photolysis of $CpFe(CO)_2X$ (1, 17, 18, 19, 20) and $CpMo(CO)_3X$ (1, 17, 19, 21, 22) has been used extensively in the preparation of new complexes, but information regarding reaction mechanisms, the nature of excited states involved, wavelength dependencies, and quantum yields has been almost totally neglected. Because of the preparative goals of most of the results reported to date, few attempts have been made to differentiate secondary or competing thermal processes. Recent results from two separate laboratories (23, 24) indicate the formation of ferrocene, $[CpFe(CO)_2]_2$, Fe^{+2}, Fe^{+3}, and Cl^- as a result of irradiating $CpFe(CO)_2Cl$ in solution.

The present report deals with our quantitative studies of the photochemical behavior of these cyclopentadienyliron– and molybdenum–carbonyl halides, including the nature of the primary photochemical processes. For the iron complexes, there is excellent agreement between our observations and the behavior predicted on the basis of the recent molecular orbital calculations of Lichtenberger and Fenske (25). Portions of the work discussed here have been published previously (26, 27).

Experimental

CpFe(CO)$_2$X and CpMo(CO)$_3$X complexes (X = Cl, Br, I) were prepared by standard literature methods (28). The linkage isomers CpFe(CO)$_2$NCS and CpFe(CO)$_2$SCN were prepared and separated according to the procedure of Sloan and Wojcicki (29). Isomerically pure *cis-* and *trans-*CpMo(CO)$_2$(PPh$_3$)X (X = Br, I) were prepared and separated by column chromatography by methods previously described (30, 31). ^{13}CO (90% enriched) was obtained from Monsanto Chemical Co. Spectral-grade solvents were used in all operations after drying by standard procedures. Samples were degassed by freeze–pump–thaw cycles or by purging with prepurified argon. Irradiations were conducted in a merry-go-round (32) of our own design. A 200 W, Hanovia medium pressure mercury arc lamp in a water-cooled immersion well surrounded by two concentric outer jackets containing appropriate filter solutions (33) supplied 366-nm or 436-nm irradiation. Light intensities (366, 4.4 × 10^{-8} ein/sec; 436, 5.5 × 10^{-8} ein/sec) were measured by ferrioxalate actinometry (34). Quantum yields were determined by monitoring UV-visible, IR, or proton NMR spectral changes, as appropriate. Conversions were limited to 15% or less and were linear with irradiation time. Reported values are the average of duplicate or triplicate determinations. Thermal reactions were negligible within the time required for photolysis and analysis, as shown by appropriate control experiments. Competition experiments involving CO or ^{13}CO were conducted with solutions that were purged with the gas just prior to photolysis.

IR spectra were obtained on Perkin Elmer 337 or 521 spectrophotometers using matched 0.1-mm KBr solution cells and cyclohexane or CCl$_4$ solvents. UV-visible and NMR measurements were carried out using Beckmann ACTA M VI and Varian T-60 instruments, respectively. A modified Varian E-12 ESR spectrometer was used for detection of radical species. Irradiations were carried out in the cavity of the instrument, monochromatic irradiation (366 or 436 nm) being achieved by using Oriel narrow band-pass glass interference filters. Mass spectra were recorded using an AEI MS-1201B spectrometer at 70-eV ionizing voltage.

Results and Discussion

Iron Complexes. The electronic spectra of CpFe(CO)$_2$X and CpFe(CO)(PPh$_3$)X complexes are dominated by an intense absorption near 350 nm which tails into the visible region. Weaker absorptions in the visible region and shoulders on the more intense band are observed in some instances. The nature of the transitions responsible are discussed later in this section. Electronic spectral data for the complexes studied

Table I. Electronic Absorption Bands for CpFe(CO)(L)X Complexes

X	L	λ_{max}, nm (ϵ, M^{-1} cm^{-1})	Solvent
Cl	CO	388(565)sh; 336(935)	C_6H_6
Br	CO	386(700)sh; 350(1028)	C_6H_6
		385(674)sh	CH_3NO_2
		385(717)sh; 346(922)	CH_3CN
I	CO	342(2090)	C_6H_6
		323(2300)	CH_3CN
NCS	CO	418(795); 340(1590); 270(4650)sh	THF
SCN	CO	525(1094); 345(1604); 278(7420)	THF
Br	PPh_3	614(160); 444(775); 360(785)sh	C_6H_6
		615(162); 440(780)	CH_3NO_2
		617(164); 440(772); 360(628)sh	CH_3CN
I	PPh_3	619(168); 440(784); 326(2510)sh	C_6H_6
		617(165); 438(770)	CH_3NO_2
		618(188); 437(781); 320(2355)sh	CH_3CN
NCS	PPh_3	550(362); 435(948); 308(2430)	THF

are given in Table I and the spectra of CpFe(CO)$_2$Cl, -Br, and -I are shown in Figure 1.

Following one-electron excitation (Reaction 7), three major chemical deactivation processes are possible for these complexes: heterolytic Fe–X cleavage (Reaction 8), homolytic Fe–X cleavage (Reaction 9), and dissociation of carbon monoxide (Reaction 10). As discussed below, we

$$\text{CpFe(CO)}_2\text{X} \xrightarrow{h\nu} \text{CpFe(CO)}_2\text{X}^* \quad \textbf{(I)} \qquad (7)$$

$$\textbf{I} \rightarrow \text{CpFe(CO)}_2^+ + \text{X}^- \qquad (8)$$

$$\textbf{I} \rightarrow \text{CpFe(CO)}_2\cdot + \text{X}\cdot \qquad (9)$$

$$\textbf{I} \rightarrow \text{CpFe(CO)X} + \text{CO} \qquad (10)$$

have found CO dissociation to be the only detectable pathway for photochemical reactions of the subject complexes at 366 or 436 nm. No net chemical reactions are observed upon irradiation of solutions of CpFe(CO)$_2$Cl, -Br, or -I with monochromatic light for periods of up to 20 min. Under 1 atm of ^{13}CO, incorporation of the isotopically labelled carbon monoxide proceeds in all cases to give CpFe(^{12}CO)(^{13}CO)X, as shown by mass spectral and IR analysis of the products. These reactions do not occur in the absence of light over 2–3 hr. IR data for the isotopically enriched species are given in Table II, along with the calculated positions of the bands attributed to CpFe(^{12}CO)(^{13}CO)X species.

Photolysis of CpFe(CO)$_2$Br or -I in benzene, tetrahydrofuran, acetonitrile, or nitromethane solutions (366 or 436 nm) in the presence of

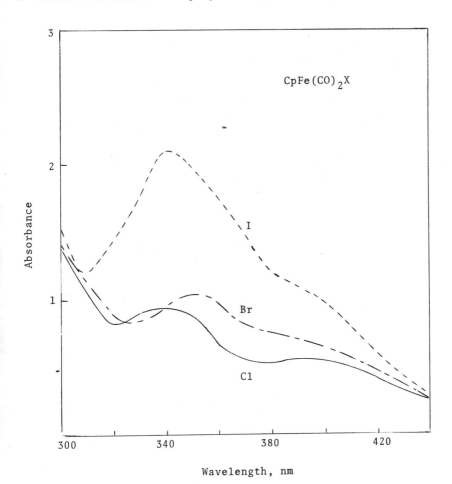

Figure 1. Electronic absorption spectra of CpFe(CO)₂Cl (———), CpFe(CO)₂-Br (— · —), and CpFe(CO)₂I (– – –) in benzene solution (1 × 10⁻³M)

Table II. Carbonyl IR Spectral Data for $CpFe(^{12}CO)_2X$ and $CpFe(^{12}CO)(^{13}CO)X$[a]

X	$CpFe(CO)_2X$[b] ν_{CO}	$CpFe(^{12}CO)(^{13}CO)X$ (obs)[b] ν_{CO}	$CpFe(^{12}CO)(^{13}CO)X$ (calc)[c] ν_{CO}
Cl	2050, 2005	2035, 1978	2037, 1973
Br	2045, 2003	2030, 1975	2035, 1972
I	2038, 2005	2025, 1968	2028, 1968

[a] Cyclohexane solution, matched 0.1-mm KBr cells.
[b] ± 2 cm⁻¹, average of three determinations.
[c] Calculated using the secular equations $\lambda_1 = \mu(k - k_i)$ and $\lambda_2 = \mu(k + k_i)$.

triphenylphosphine leads to efficient photosubstitution (Reaction 11). There is no thermal reaction during the time of the photolysis, as determined by appropriate thermal control reactions. Unfortunately, the

$$CpFe(CO)_2X + PPh_3 \xrightarrow[(X = Br, I)]{h\nu} CpFe(CO)(PPh_3)X + CO \quad (11)$$

chloro derivative reacts rapidly with triphenylphosphine in a variety of solvents at room temperature to afford the ionic derivative [CpFe(CO)$_2$-PPh$_3$]Cl (*17*), prohibiting investigation of its photochemical reactions. Quantum yields for the observed photosubstitution reactions of the bromo and iodo complexes are given in Table III. The quantum yields for triphenylphosphine substitution are decreased dramatically by addition of carbon monoxide (1 atm), as shown in Table III. Taken together with our observations concerning ^{13}CO incorporation, this strongly suggests the intermediacy of CpFe(CO)X subsequent to excitation (Reac-

Table III. Quantum Yields for the Reactions
CpFe(CO)$_2$X + PPh$_3$ → CpFe(CO)(PPh$_3$)X + CO[a,b]

X	Irradiation λ (nm)	Solvent	Φ(±10%)
Br	366	C$_6$H$_6$	0.72
Br	366	C$_6$H$_6$/CO[c]	0.18
Br	366	CH$_3$CN	0.89
Br	366	CH$_3$CN/NaBPh$_4$	0.86
Br	366	CH$_3$CN/C$_6$H$_6$O$_2$[d]	0.75
Br	436	C$_6$H$_6$	0.82
Br	436	C$_6$H$_6$/CO[c]	0.58
Br	436	CH$_3$NO$_2$	0.40
Br	436	CH$_3$NO$_2$/NaBPh$_4$	0.46
Br	436	CH$_3$NO$_2$/C$_6$H$_6$O$_2$[d]	0.34
I	366	C$_6$H$_6$	0.63
I	366	C$_6$H$_6$/CO[c]	0.14
I	366	CH$_3$CN	0.21
I	366	CH$_3$CN/NaBPh$_4$	0.19
I	366	CH$_3$CN/C$_6$H$_6$O$_2$[d]	0.23
I	436	C$_6$H$_6$	0.69
I	436	C$_6$H$_6$/CO[c]	0.40
I	436	CH$_3$NO$_2$	0.40
I	436	CH$_3$NO$_2$/NaBPh$_4$	0.48
I	436	CH$_3$NO$_2$/C$_6$H$_6$O$_2$[d]	0.38

[a] Determined by appearance of long wavelength absorption band of CpFe(CO)(PPh$_3$)X.
[b] At 366 nm, [CpFe(CO)$_2$X] = [PPh$_3$] = 1 × 10^{-3}M. At 436 nm, [CpFe(CO)$_2$X] = [PPh$_3$] = 2 × 10^{-3}M.
[c] One atm CO pressure (*see* "Experimental" section).
[d] C$_6$H$_6$O$_2$ = hydroquinone.

tion 10). Further support for this hypothesis comes from a study of quantum-yield dependence upon PPh_3 concentration (Figure 2). When the concentration of $CpFe(CO)_2Br$ is greater than $[PPh_3]$, relatively low values of Φ are observed. The quantum yields increase smoothly with increasing $[PPh_3]$ and reach a limiting value above which additional ligand has no effect on Φ. Similar data are obtained for $CpFe(CO)_2I$, and this phenomenon does not appear to depend upon either wavelength or solvent (27). This behavior is precisely what is expected for a dissociative process in which the incoming nucleophile must capture low concentrations of a coordinatively unsaturated intermediate (35).

The absence of the ionic derivatives $[CpFe(CO)_2L]X$ ($L = PPh_3$ or CO) among the photoproducts when irradiations are conducted in the presence of the added ligand constitute convincing, although negative, evidence that heterolytic Fe–X cleavage (Reaction 8) does not occur. The $CpFe(CO)_2^+$ or $CpFe(CO)_2(S)^+$ intermediate (S = solvent) generated by this process has been shown to react very rapidly with nucleophiles such as CO or PPh_3 (36). More compelling in this regard is our observation that quantum yields for substitution are not affected to a significant extent by changing the solvent from benzene to the more polar acetonitrile or nitromethane or by the presence of added electrolytes (Table III). Furthermore, we find that irradiation of saturated solutions of $[CpFe(CO)_2(PPh_3)]X$ in benzene or nitromethane does not lead to the observed photoproduct $CpFe(CO)(PPh_3)X$, precluding the intermediacy of the ionic products in the overall reaction scheme.

Homolytic photochemical cleavage of the metal–metal bonds in dimeric metal carbonyls has been demonstrated for $[CpMo(CO)_3]_2$, $[CpW(CO)_3]_2$, $Mn_2(CO)_{10}$, and $Re_2(CO)_{10}$ (10, 11, 12, 13, 14), but to our knowledge, homolyses of metal–halogen bonds are unknown at present. Nevertheless we conducted experiments designed to demonstrate whether such processes were occurring in the systems under investigation here. Quantum yields for the photosubstitution processes are unaffected by the addition of the radical scavenger hydroquinone to the photolyte, indicating that metal- or halogen-based radicals are not involved in the substitution process (14). The radical species $CpM(CO)_3\cdot$ (M = Mo or W) and $(CO)_5M\cdot$ (M = Mn or Re) (10, 13) react rapidly with carbon tetrachloride to give the corresponding metal chloride. Reasoning that the species $CpFe(CO)_2\cdot$ should behave similarly (24), we irradiated all combinations of $CpFe(CO)_2X$ and CX'_4 in benzene solution (X = Cl, Br, I; X' = Cl or Br). We found no evidence (IR, MS) for formation of $CpFe(CO)_2X'$ in any of these experiments. Irradiation of benzene solutions of the iron complexes with added Ph_3CCl in the cavity of an ESR spectrometer did not produce the spectrum of the trityl radical, once again indicating that $CpFe(CO)_2\cdot$ is not formed. In

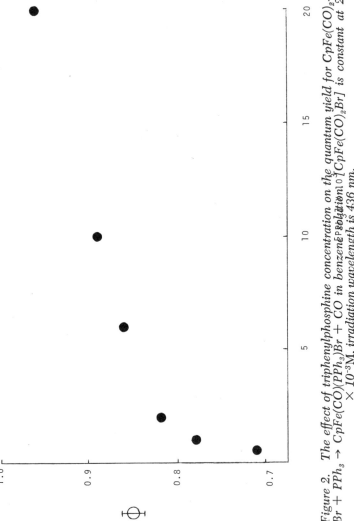

Figure 2. The effect of triphenylphosphine concentration on the quantum yield for $CpFe(CO)_2Br + PPh_3 \rightarrow CpFe(CO)(PPh_3)Br + CO$ in benzene solution. $[CpFe(CO)_2Br]$ is constant at $2 \times 10^{-3}M$, irradiation wavelength is 436 nm.

accord with literature reports (*10, 13*), we observed formation of the trityl radical by photolysis of [CpMo(CO)$_3$]$_2$ or Mn$_2$(CO)$_{10}$ in the presence of Ph$_3$CCl under these conditions. Taken together, these observations allow the elimination of radical Fe–X bond scission as an important photochemical pathway for the CpFe(CO)$_2$X complexes under the conditions used here.

The recent molecular orbital calculations of Lichtenberger and Fenske (*25*) provide a rational basis for interpretation of the observed photochemical behavior of the CpFe(CO)$_2$X complexes. The highest-occupied MOs for these molecules are a doubly degenerate set of iron–halogen π^* orbitals, as shown in Figure 3 for CpFe(CO)$_2$Cl. The calculations are wholly consistent with the observed photoelectron spectra of CpFe(CO)$_2$Cl, -Br, and -I (*25*). It is reasonable to assume

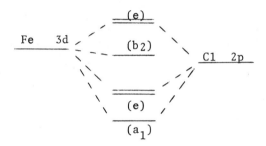

Journal of the American Chemical Society

Figure 3. Highest-filled molecular orbitals for CpFe(CO)$_2$Cl. Assignments in order of decreasing energy: e = Fe–Cl π^*; b$_2$ = Fe nonbonding; e = Fe–Cl π, a$_1$ = FeCl σ. All orbitals are completely filled (25).

that absorption of light by these complexes is the result of promotion from the highest-filled orbital(s). The effect of such a transition, regardless of the orbital to which promotion occurs, is to increase the Fe–X bond order, since an antibonding orbital is depopulated. Thus any photochemical process involving rupture of the Fe–X bond is predicted to be highly unlikely, in accord with our observations. A more complete understanding of the photochemical behavior of these complexes requires information regarding the nature of low-lying excited states and their lifetimes. Further investigations designed to elucidate such information are currently in progress.

Our interest in the photochemistry of the subject iron complexes was originated by the observation that the linkage isomers CpFe(CO)$_2$-NCS and CpFe(CO)$_2$SCN (*29*) are readily interconverted photo-

$$\text{CpFe(CO)}_2\text{NCS} \xrightleftharpoons{h\nu} \text{CpFe(CO)}_2\text{SCN} \qquad (12)$$

chemically in THF solution (26). The IR spectra of irradiated solutions of either isomer (Figure 4) clearly show the isomerization process, which does not occur thermally in solution at room temperature (26, 29). We have been unable to obtain quantum yield data for these processes because of the production of an unidentified product that absorbs a significant fraction of the incident light over the 350–500 nm range. Repeated attempts to isolate this product have failed. Several features of the photochemistry of these species deserve comment, however, as they reinforce and expand on the conclusions reached for the halide derivatives. The isomerization of CpFe(CO)$_2$NCS to CpFe(CO)$_2$SCN is totally inhibited by carbon monoxide (1 atm), and when ^{13}CO is used, CpFe(^{12}CO)(^{13}CO)NCS is produced. Photolysis of the N-bonded isomer in the presence of triphenylphosphine affords CpFe(CO)(PPh$_3$)NCS with no evidence of linkage isomerization. These observations parallel those made for the halide complexes above and strongly suggest that CO

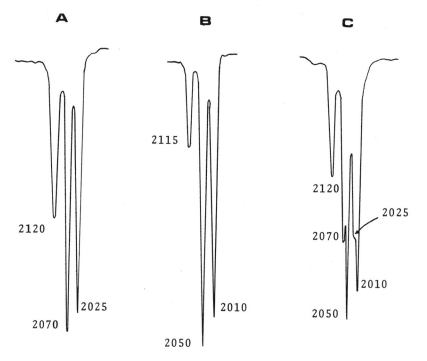

Figure 4. IR spectra of: (A) CpFe(CO)$_2$NCS, (B) CpFe(CO)$_2$SCN, and (C) the isomeric mixture at the photostationary state (436 nm). Irradiations conducted in THF solution, spectra recorded in chloroform.

dissociation is the primary photochemical pathway. The photoisomerization of CpFe(CO)$_2$SCN is slowed but not totally inhibited by added ligands (CO or PPh$_3$). On the basis of the data at hand, we favor a mechanism for the isomerization involving: (1) loss of CO, (2) intramolecular "turnaround" of thiocyanate, and (3) thermal recapture of CO to give either isomer. The failure of added ligands to quench the -SCN → NCS isomerization process is likely a consequence of the expected bond angles Fe–S–C ($\sim 105°$) and Fe–N–C ($\sim 180°$) (37). After photolytic CO loss, the sulfur-bonded form is very near the geometry required for the "turn-around" process, and attack of an external ligand (CO or PPh$_3$) cannot effectively compete for the unsaturated CpFe(CO)SCN intermediate. The expected (37) linearity of the Fe–N–C–S linkage in the corresponding CpFe(CO)NCS species should allow

Table IV. Electronic Spectral Data for CpMo(CO)$_n$(PPh$_3$)$_{3-n}$X Complexes (Benzene Solution) [a]

Complex	λ_{max}, nm (ϵ, M^{-1} cm^{-1})
CpMo(CO)$_3$Cl	320(1980)sh, 470(504)
cis-CpMo(CO)$_2$(PPh$_3$)Cl	360(936)sh, 473(689)
CpMo(CO)(PPh$_3$)$_2$Cl	[b] 467(604)
CpMo(CO)$_3$Br	[b] 475(471)
cis-CpMo(CO)$_2$(PPh$_3$)Br	365(820)sh, 475(488)
trans-CpMo(CO)$_2$(PPh$_3$)Br	314(2210)sh, 470(252)
CpMo(CO)$_3$I	315(3810), 482(560)
cis-CpMo(CO)$_2$(PPh$_3$)I	300(4260)sh, 484(562)
trans-CpMo(CO)$_2$(PPh$_3$)I	308(5960), 475(567)
CpMo(CO)(PPh$_3$)$_2$I	284(5620), 470(1112)

[a] All complexes exhibit a further absorption near 250 nm which is obscured by this solvent.
[b] Absorption in this region not resolved for these complexes.

for a more effective competition of added ligands, as observed. These arguments must remain speculative in the absence of quantum yield data for the photo isomerizations, but the rarity of such processes merits the inclusion of our observations in the current discussion.

Molybdenum Complexes. The molybdenum derivatives CpMo(CO)$_3$X, CpMo(CO)$_2$(L)X, and CpMo(CO)(L)$_2$X exhibit distorted square pyramidal geometries (η^5-C$_5$H$_5$ in the apical position, in solution, and in the solid state (30, 31).) These complexes thus provide an opportunity to study the geometrical consequences of photosubstitution. Regardless of the degree of substitution, all of these complexes exhibit similar electronic spectra (Table IV) consisting of a strong, near UV absorption ($\lambda_{max} \cong 325$ nm) and a much weaker visible absorption ($\lambda_{max} \cong 475$ nm). Lacking a theoretical study of these complexes

Table V. Quantum Yields for the Reaction $CpMo(CO)_3X + PPh_3 \rightarrow cis\text{-}CpMo(CO)_2(PPh_3)X$ [a]

(Benzene solution)

X	Irradiation λ (nm)	Φ (±10%)
Br	366	0.74
Br	436	0.37
I	366	0.32
I	436	0.16

[a] Monitored by appearance of product resonance in proton NMR spectrum (30, 31).
[b] $[CpMo(CO)_3X] = [PPh_3] = 1 \times 10^{-3}M$ (366 nm); $[CpMo(CO)_3X] = [PPh_3] = 2 \times 10^{-3}M$ (436 nm).

comparable with that for the iron series (25), the origins of these transitions are not clear at present.

Irradiation of the various complexes at either 366 or 436 nm lead to the processes summarized in Reactions 13, 14, 15, 16, and 17. Quantum yield data for the bromo and iodo complexes are given in Tables V and VI. The thermal ligand exchange processes of $CpMo(CO)_3Cl$ (Reaction 13) and the corresponding mono- and disubstituted derivatives (Reactions 15 and 16) are very facile at 25°C in solution, prohibiting analysis of this system.

$$CpMo(CO)_3X + PPh_3 \rightarrow cis\text{-}CpMo(CO)_2(PPh_3)X + CO \quad (13)$$

$$cis\text{-}CpMo(CO)_2(PPh_3)X \rightleftarrows trans\text{-}CpMo(CO)_2(PPh_3)X \quad (14)$$

Table VI. Quantum Yields for Reactions of $CpMo(CO)_2(PPh_3)X$ (Benzene Solution) [a,b]

Complex	Irradiation λ (nm)	Φ_{14} [c]	Φ_{15} [c]	Φ_{16} [c]
$cis\text{-}CpMo(CO)_2(PPh_3)Br$	366	0.036	0.072	0.14
$cis\text{-}CpMo(CO)_2(PPh_3)Br$	436	0.018	0.066	0.14
$trans\text{-}CpMo(CO)_2(PPh_3)Br$	366	0.40	0.38	0.26
$trans\text{-}CpMo(CO)_2(PPh_3)Br$	436	0.23	0.11	0.16
$cis\text{-}CpMo(CO)_2(PPh_3)I$	366	0.17	0.21	0.086
$cis\text{-}CpMo(CO)_2(PPh_3)I$	436	0.027	0.031	0.081
$trans\text{-}CpMo(CO)_2(PPh_3)I$	366	0.12	0.086	0.12
$trans\text{-}CpMo(CO)_2(PPh_3)I$	436	0.021	0.004	0.032

[a] Determined by disappearance and appearance, respectively, of starting material and product resonances in the C_5H_5 region of the proton NMR spectrum.
[b] All concentrations $= 1 \times 10^{-3}M$.
[c] Subscripts refer to the corresponding equations in the text.

cis- or $trans$-CpMo(CO)$_2$(PPh$_3$)X + PPh$_3$ →
$trans$-CpMo(CO)(PPh$_3$)$_2$X + CO   (15)

cis- or $trans$-CpMo(CO)$_2$(PPh$_3$)X + CO →
CpMo(CO)$_3$X + PPh$_3$ (16)

$trans$-CpMo(CO)(PPh$_3$)$_2$I + CO →
cis-CpMo(CO)$_2$(PPh$_3$)I + PPh$_3$ (17)

Triphenylphosphine substitution into CpMo(CO)$_3$X could, in principle, lead to either of two covalent products, cis- or $trans$-CpMo(CO)$_2$-(PPh$_3$)X, or a mixture thereof. We observe exclusive formation of the cis isomer (Reaction 13, Table V) when conversions of CpMo(CO)$_3$X

```
    ⌢              ⌢
    |              |
OC--M--CO      OC--M--X
   ╱ ╲             ╱ ╲
  L   X           L   CO
   cis            trans
```

are held to 30% or less. Figure 5 provides a framework for discussing the observed stereoselectivity of product formation. Loss of a carbon monoxide initially cis or trans to halogen would lead to the coordinatively unsaturated intermediates I or II, respectively. Interconversion of these species is possible via rearrangement to III. As shown in Figure 5, I leads to the cis isomer upon addition of phosphine, II leads to the trans isomer, and III should afford both isomers. In the ground state, both IR (38) and carbon-13 (39) data indicate that the carbonyl trans to halogen is back-bonded to molybdenum to a greater extent than are the two cis carbonyls. Insofar as greater back-bonding reflects greater bond strength, we would then expect preferential loss of CO cis to the halogen if these relative bond-strength differences persist in the excited state. Exclusive formation of cis-CpMo(CO)$_2$(PPh$_3$)X is consistent with selective loss of CO cis to halogen and subsequent capture of intermediate I by the phosphorus ligand. This result should obtain even in the presence of rapid I ⇌ III ⇌ II equilibration since, by the principle of microscopic reversibility (40, 41), the entering ligand must occupy the position vacated by the ligand that dissociated originally. A cautionary note regarding the application of microscopic reversibility arguments to processes in which one step is photochemical and the reverse thermal in nature has been given by Darensbourg, Nelson, and Murphy (41). We fully agree with these authors and point out the additional dangers of such an approach when the leaving and entering groups (in this case CO and PPh$_3$, respectively) are nonidentical. We are presently engaged in a study of both thermal and photochemical ligand exchange processes of the subject complexes designed to test the above hypothesis of selective labilization.

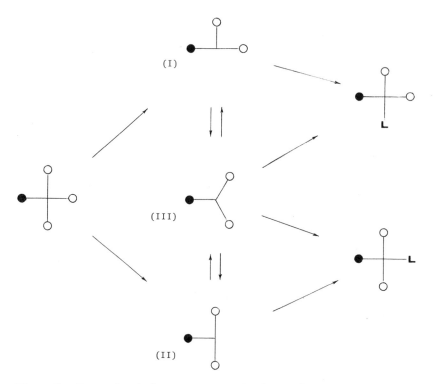

Figure 5. Stereochemical consequences of selective loss of CO from CpMo(CO)$_3$X in the formation of CpMo(CO)$_2$(PPh$_3$)X, including possibility of rearrangement of intermediates. The cyclopentadienyl ligand has been omitted for clarity. (●) = halide; (○) = CO, L = PPh$_3$.

Geometric isomerization of cis- and trans-CpMo(CO)$_2$(PPh$_3$)X (X = Br or I) occurs on photolysis at either 366 or 436 nm (Reaction 14). Our initial observations (26) for these systems showed that prolonged irradiation led to formation of small amounts of CpMo(CO)$_3$X and trans-CpMo(CO)(PPh$_3$)$_2$X as well as the cis/trans isomer ratio characteristic of the photostationary state. These "disproportionation" products seemingly must arise by competitive loss of CO and PPh$_3$ from the starting material to give the coordinatively unsaturated intermediates

CpMo(CO)$_2$X and CpMo(CO)(PPh$_3$)X, respectively (Reactions 18 and 19). Recombination of these fragments with a ligand other than that

$$CpMo(CO)_2(PPh_3)X \rightarrow CpMo(CO)_2X + PPh_3 \qquad (18)$$

$$CpMo(CO)_2(PPh_3)X \rightarrow CpMo(CO)(PPh_3)X + CO \qquad (19)$$

which dissociated (i.e., CO vs. PPh$_3$) would lead to the observed products CpMo(CO)$_3$X and CpMo(CO)(PPh$_3$)$_2$X. Aside from determining the origin of these products, a principal concern is whether the isomerization process is dissociative or intramolecular. The latter pathway is well established in the thermal reactions of these and structurally related complexes (42, 43).

Quantum yield data for the reactions given by Reactions 14, 15, and 16 (X = Br or I) are given in Table VI. In the absence of added ligands, cis–trans isomerization is the only process observed, conversions being limited to 10% or less. Addition of triphenylphosphine or carbon monoxide (1 atm) does not significantly affect the observed isomerization quantum yields but does lead to the CpMo(CO)(PPh$_3$)$_2$X and CpMo(CO)$_3$X products, respectively. For example, the isomerization quantum yield for cis-CpMo(CO)$_2$(PPh$_3$)I, Φ_{14}, is 0.17 at 366 nm in the absence of added ligand (Table VI.) The disappearance quantum yield for this complex in the presence of added triphenylphosphine is the sum of Φ_{14} and Φ_{15}, 0.38. Similarly, under 1 atm of CO, $\Phi_{14} = 0.17$ and $\Phi_{16} = 0.086$, and the disappearance quantum yield equals the sum for the two processes. If ligand dissociation were involved in the isomerization, Φ_{14} (the isomerization quantum yield) should be decreased by added ligands, contrary to our observations. On the basis of these results, we suggest that the isomerization process does not involve dissociation of either CO or the phosphorus ligand. If this is actually the case, these complexes have three distinct excited state deactivation pathways available, i.e., those given by Reactions 14, 18, and 19. The dissociative processes can lead to a variety of intermediates depending upon the geometry of the starting complex and which ligand is lost. One piece of information currently in hand suggests that we will be able to definitively map the course of these reactions. *Trans*-CpMo(CO)(PPh$_3$)$_2$I produces *cis*-CpMo(CO)$_2$(PPh$_3$)I but not the trans isomer when photolyzed under 1 atm of CO. As in the case of the stereospecific substitution of the CpMo(CO)$_3$X complexes discussed above, this observation is consistent with a coordinatively unsaturated intermediate having appropriate stereochemistry for production of one and only one product. We have embarked on a series of investigations designed to answer the intriguing questions raised by these preliminary results.

Conclusion

The photochemistry of the complexes described here is dominated by loss of carbon monoxide or other two-electron ligands, with radical or ionic processes being of little or no importance as excited state decay pathways. Although labilization of the cyclopentadienyl ligand is generally thought to be unlikely as a result of one-electron excitation (1), the recently reported ring-exchange reactions between $(\eta^5\text{-}C_5H_5)_2ZrCl_2$ and $(\eta^5\text{-}C_5D_5)_2ZrCl_2$ (44) indicate that this possibility should be more closely examined. The photochemically induced geometric and linkage isomerizations observed in the present study may well prove to be of synthetic utility in these and related organometallic systems, and this possibility is being investigated.

Acknowledgment

We wish to thank R. P. Stewart for carrying out the calculations on the IR-stretching force constants and frequencies and D. J. Darensbourg, T. L. Brown, and R. W. Murray for valuable discussions. The assistance of R. H. Ahmed in carrying out the ESR measurements is gratefully acknowledged. This research was supported in part by the Office of Research, University of Missouri–St. Louis, in the form of a research assistantship to DGA.

Literature Cited

1. Wrighton, M. S., *Chem. Rev.* (1974) **74**, 401.
2. von Gustorf, E. K., Grevels, F. W., *Fortschr. Chem. Forsch.* (1969) **13**, 366.
3. Balzani, V., Carassiti, V., "Photochemistry of Coordination Compounds," Academic, New York, 1970.
4. Stolz, I. W., Dobson, G. R., Sheline, R. K., *J. Am. Chem. Soc.* (1963) **85**, 1013.
5. Burdett, J. K., Graham, M. A., Perutz, R. N., Poliakoff, M., Rest, A. J., Turner, J. J., Turner, R. F., *J. Am. Chem. Soc.* (1975) **97**, 4805.
6. Perutz, R. N., Turner, J. J., *J. Am. Chem. Soc.* (1975) **97**, 4800.
7. Kelly, J. M., Hermann, H., von Gustorf, E. K., *J. Chem. Soc., Chem. Commun.* (1973) 105.
8. Wrighton, M., *Inorg. Chem.* (1974) **13**, 905.
9. Wrighton, M., Hammond, G. S., Gray, H. B., *Mol. Photochem.* (1973) **5**, 179.
10. Ginley, D. S., Wrighton, M. S., *J. Am. Chem. Soc.* (1975) **97**, 4908.
11. Hughey, J. L., IV, Bock, C. R., Meyer, T. J., *J. Am. Chem. Soc.* (1975) **97**, 4440.
12. Hughey, J. L., IV, Anderson, C. P., Meyer, T. J., *J. Organomet. Chem.* (1977) **125**, C49.
13. Wrighton, M. S., Ginley, D. S., *J. Am. Chem. Soc.* (1975) **97**, 4246.
14. Byers, B. H., Brown, T. L., *J. Am. Chem. Soc.* (1977) **99**, 2527.
15. Giordano, P. J., Wrighton, M. S., *Inorg. Chem.* (1977) **16**, 160.

16. Wrighton, M. S., Morse, D. L., Gray, H. B., Otteson, D. K., *J. Am. Chem. Soc.* (1976) **98**, 1111.
17. Treichel, P. M., Shubkin, R. L., Barnett, K. W., Reichard, D., *Inorg. Chem.* (1966) **5**, 1177.
18. King, R. B., Kapoor, P. N., Kapoor, R. N., *Inorg. Chem.* (1971) **10**, 1841.
19. King, R. B., Kapoor, R. N., Saran, M. S., Kapoor, P. N., *Inorg. Chem.* (1971) **10**, 1851.
20. King, R. B., Zipperer, W. C., Ishaq, M., *Inorg. Chem.* (1972) **11**, 1361.
21. Haines, R. J., Nyholm, R. S., Stiddard, M. H. B., *J. Chem. Soc.* A (1966) 1606.
22. Barnett, K. W., Treichel, P. M., *Inorg. Chem.* (1967) **6**, 294.
23. Ali, L. H.. Cox, A., Kemp, T. J., *J. Chem. Soc.. Dalton Trans.* (1973) 1475.
24. Giannotti, C., Merle, G., *J. Organomet. Chem.* (1976) **105**, 97.
25. Lichtenberger, D. L., Fenske, R. F., *J. Am. Chem. Soc.* (1976) **98**, 50.
26. Alway, D. G., Barnett, K. W., *J. Organomet. Chem.* (1975) **99**, C52.
27. Alway, D. G., Barnett, K. W., *Inorg. Chem.*, in press.
28. King, R. B., "Organometallic Syntheses," Vol. I, Academic, New York, 1965.
29. Sloan, T. E., Wojcicki, A., *Inorg. Chem.* (1968) **7**, 1268.
30. Beach, D. L., Barnett, K. W., *J. Organomet. Chem.* (1975) **97**, C27.
31. Beach, D. L., Dattilo, M., Barnett, K. W., *J. Organomet. Chem.* (1977) **140**, 47.
32. Moses, F. G., Liu, R. S. H., Monroe, B. M., *Mol. Photochem.* (1969) **1**, 245.
33. Calvert, J. G., Pitts, J. N., "Photochemistry," Wiley, New York, 1966.
34. Hatchard, C. G., Parker, C. A., *Proc. R. Soc. London, Ser.* A (1956) **235**, 518.
35. Basolo, F., Pearson, R. G., "Mechanisms of Inorganic Reactions," 2nd ed., Wiley, New York, 1967.
36. Ferguson, J. A., Meyer, T. J., *Inorg. Chem.* (1971) **10**, 1025.
37. Norbury, A. H., *Adv. Inorg. Chem. Radiochem.* (1975) **17**, 231.
38. King, R. B., Houk, L. W., *Can. J. Chem.* (1969) **47**, 2959.
39. Todd, L. J., Wilkinson, J. R., Hickey, J. P., Beach, D. L., Barnett, K. W., *J. Organomet. Chem.* (1978) in press.
40. Atwood, J. D., Brown, T. L., *J. Am. Chem. Soc.* (1976) **98**, 3155, 3160, and references therein.
41. Darensbourg, D. J., Nelson, H. H., III, Murphy, M. A., *J. Am. Chem. Soc.* (1977) **99**, 896.
42. Faller, J. W., Anderson, A. S., *J. Am. Chem. Soc.* (1970) **92**, 5852.
43. Flood, T. C., Rosenberg, E., Sarhangi, A., *J. Am. Chem. Soc.* (1977) **99**, 4334.
44. Peng, H. M., Brubaker, C. H., *J. Organomet. Chem.* (1977) **135**, 333.

RECEIVED September 20, 1977.

8

Photo-Induced Declusterification of $HCCo_3$-$(CO)_9$, $CH_3CCo_3(CO)_9$, and $HFeCo_3(CO)_{12}$

GREGORY L. GEOFFROY[1] and RONALD A. EPSTEIN

Department of Chemistry, Pennsylvania State University, University Park, PA 16802

Irradiation of solutions of $HCCo_3(CO)_9$ under a hydrogen atmosphere with visible or UV light leads to quantitative formation of $Co_4(CO)_{12}$ and production of methane. Under a 3:1 H_2:CO atmosphere, irradiation produces $Co_2(CO)_8$ with a 366-nm quantum yield of 0.03. Photolysis in the presence of D_2 has shown that the methane derives from the apical CH group and not from CO. Irradiation of $HCCo_3$-$(CO)_9$ in the presence of hydrogen and 1-hexene leads to catalytic isomerization to cis- and trans-2-hexene. The photochemical properties of $CH_3CCo_3(CO)_9$ parallel those of $HCCo_3(CO)_9$ except that no reaction is observed under a hydrogen–carbon monoxide atmosphere. Irradiation of $HFeCo_3(CO)_{12}$ and $HFeCo_3(CO)_{10}(PPh_3)_2$ in degassed solutions also leads to declusterification with subsequent formation of $Co_4(CO)_{12}$ and $Co_2(CO)_6(PPh_3)_2$, respectively. The iron-containing products were not identified.

Numerous organometallic cluster complexes have been prepared and studied in recent years, and the thermal reactivity of a large number of these is well documented (*1, 2, 3*). The photochemical properties of organometallic clusters, however, have not been studied in detail although a few interesting reports have appeared (*4–17*). Since photolysis of monomeric complexes often leads to very active species of both synthetic and catalytic utility, it is likely that photolysis of clusters may yield similar results. For this reason we have set out to explore the photochemistry of selected transition metal cluster complexes in detail and aim to develop an understanding of those factors that control cluster photoreactivity.

[1] Author to whom correspondence should be sent.

It is appropriate to review briefly here those photochemical studies that have been reported. Perhaps the most definitive work to date has been that of Lewis and co-workers (5, 6) who showed that photolysis of $Ru_3(CO)_{12}$ in the presence of CO, C_2H_4, and PPh_3 led to declusterification and the monomeric products shown in Reactions 1, 2, and 3.

$$Ru_3(CO)_{12} \xrightarrow[CO]{h\nu} [Ru(CO)_5] \ (100\%) \quad (1)$$

$$Ru_3(CO)_{12} \xrightarrow[C_2H_4]{h\nu} [Ru(CO)_4(C_2H_4)] \quad (2)$$

$$Ru_3(CO)_{12} \xrightarrow[PPh_3]{h\nu} [Ru(CO)_4PPh_3] + \quad (3)$$
$$[Ru(CO)_3(PPh_3)_2]$$

The reaction was subsequently extended by other workers to the preparation of cyclooctatetraene (COT) and olefin derivatives, Reactions 4 and 5 (7, 8).

$$Ru_3(CO)_{12} \xrightarrow[olefin]{h\nu} [Ru(CO)_4(olefin)] \quad (4)$$
(olefin = ethylacrylate, diethylfumarate)

$$Ru_3(CO)_{12} \xrightarrow[COT]{h\nu} [Ru(CO)_3(COT)] \quad (5)$$

A similar reaction with COT and $Os_3(CO)_{12}$ was obtained, Reaction 6 (9).

$$Os_3(CO)_{12} \xrightarrow[COT]{h\nu, \ 10 \ days} [Os(CO)_3(COT)] \quad (6)$$

Stone and co-workers (10, 11, 12, 13) reported that irradiation of $Ru_3(CO)_{12}$ in the presence of trialkylsilanes and -germanes led to declusterification and to the products shown in Reaction 7.

$$Ru_3(CO)_{12} + (CH_3)_3MH \xrightarrow{h\nu} [Ru(CO)_4\{M(CH_3)_3\}_2] \quad (7)$$
$$M = Si, Ge \qquad\qquad + [Ru(CO)_4\{M(CH_3)_3\}]_2$$

Spectroscopic evidence indicated that the $[Ru(CO)_4\{M(CH_3)_3\}]_2$ dimer possesses a linear M–Ru–Ru–M structure (*10*). Prolonged irradiation (five weeks) of $Os_3(CO)_{12}$ in the presence of $(CH_3)_3MH$ (M=Si, Ge, Sn) gave the analogous osmium derivatives (*11*). Although the overall mechanisms of these reactions have not been worked out, it is likely that photolysis simply leads to cleavage of the metal trimers into $M(CO)_4$ units which subsequently add CO, PPh_3, and olefin, or undergo oxidative addition of R_3MH and subsequent elimination of hydrogen, to give the products shown. Quantum yields have not been measured, but the relative data indicate that it is much more difficult to cleave $Os_3(CO)_{12}$ than $Ru_3(CO)_{12}$. Photolysis of $Fe_3(CO)_{12}$ and $Ru_3(CO)_{12}$ in the presence of phosphine- or arsine-substituted fluoroolefins such as Structure 1 has been

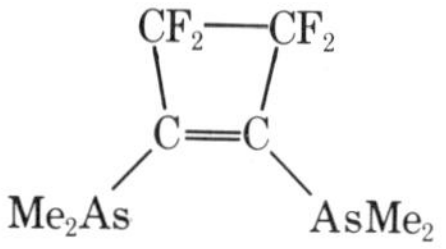

$$\begin{array}{c} CF_2 \text{---} CF_2 \\ | \quad\quad | \\ C = C \\ Me_2As \quad\quad AsMe_2 \end{array}$$

1

reported to lead to iron and ruthenium monomers, dimers, and trimers (*14, 15, 16*). Finally, irradiation of $H_4Os_4(CO)_{12}$ in the presence of CH_2CHR (R = H, Ph, ^{t}Bu) gave formation of $H_3Os_4(CO)_{11}(CHCHR)$ derivatives which were proposed to have the structure shown in 2 (*17*), and irradiation of $Fe_4(\eta^5-C_5H_5)_4(CO)_4$ in CCl_4 solution gave oxidation of the cluster to its monocation, Reaction 8 (*18*).

$$Fe_4(\eta^5-C_5H_5)_4(CO)_4 \xrightarrow[CCl_4]{h\nu} [Fe_4(\eta^5-C_5H_5)_4(CO)_4]Cl \quad\quad (8)$$

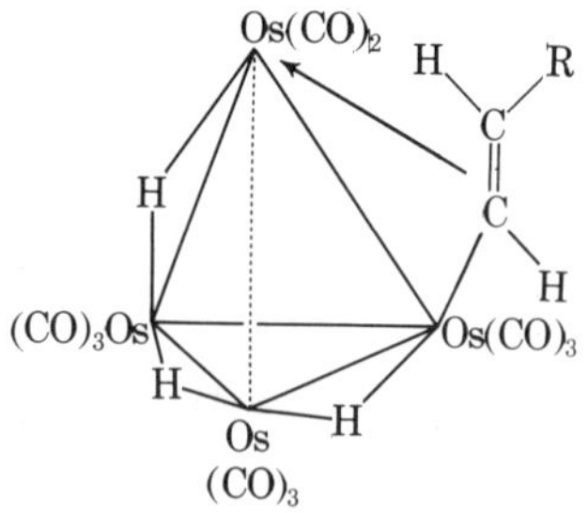

2

The first class of compounds that we have chosen to examine are the alkylidynetricobaltnonacarbonyl clusters, $YCCo_3(CO)_9$. A very large number of such clusters is known (*19, 20, 21*), and Y can vary through a wide range of substituent groups. The structure of $CH_3CCo_3(CO)_9$ has been determined by x-ray diffraction (22) and is depicted in Structure 3.

$$\begin{array}{c} CH_3 \\ | \\ C \\ OC \diagdown \diagup \diagdown CO \\ OC-Co \diagup\!\!\!\diagdown Co-CO \\ OC \diagup \; Co \; \diagdown CO \\ \diagup | \diagdown \\ OC \; CO \; CO \end{array}$$

3

The core of this cluster consists of a triangle of metal–metal bonded cobalt atoms with a symmetrical triply bridging carbon atom forming the apex of a trigonal pyramid. The carbonyl groups are terminal in this example, although bridging carbonyls are found in some substituted $YCCo_3(CO)_{9-x}L_x$ clusters (23). The chemistry of this class of compounds has been explored in great detail although the majority of studies has centered on modifications of the substituent Y (19, 20, 21). Studies of the chemistry of the Co_3C portion have focused mainly on reactions in which CO is replaced by other ligands, principally phosphines and arsines (19, 20, 21, 23).

We report herein the spectral and photochemical properties of $HCCo_3(CO)_9$ and $CH_3CCo_3(CO)_9$, the two simplest members of this class of compounds. These are compared with the structurally similar $HFeCo_3(CO)_{12}$ and $HFeCo_3(CO)_{10}(PPh_3)_2$, where in effect an $Fe(CO)_3$ unit has replaced the YC group, and the hydride is believed to bridge the opposite face of the Co_3 triangle (24).

Experimental Section

The complexes $HCCo_3(CO)_9$ (25), $CH_3CCo_3(CO)_9$ (26), $CH_3CCo_3(CO)_8PPh_3$ (23), $CH_3CCo_3(CO)_7(PPh_3)_2$ (23), $HFeCo_3(CO)_{12}$ (27), $HFeCo_3(CO)_{10}(PPh_3)_2$ (28), $Co_4(CO)_{12}$ (29), $Na[Co_3(CO)_{10}]$ (30), and $Co_2(CO)_6(PPh_3)_2$ (31) were prepared by published procedures. All solvents were distilled from $LiAlH_4$ and degassed on a vacuum line prior to use. A 3:1 H_2:CO gas mixture was obtained from Matheson Gas Products and was used without further purification.

General Irradiation Procedures. Solutions to be irradiated were prepared on a vacuum line using specially constructed degassable quartz UV cells or Schlenk vessels. After heating the evacuated vessel to remove traces of adsorbed water, the compound to be studied was added and the vessel reevacuated. Because of the volatility of $HCCo_3(CO)_9$ and $CH_3CCo_3(CO)_9$, it was necessary to cool these samples in liquid nitrogen during evacuation. Dry and degassed solvent was then distilled into the vessel and the appropriate atmosphere placed over the sample. Irradiations were conducted at room temperature (21°–25°C) using a 450-W

Hanovia medium-pressure Hg lamp equipped with the appropriate Corning glass filters ($\lambda \geq 500$ nm, Corning #3-69; $\lambda = 366$ nm, Corning #7-83 narrow bandpass filter), a 100-W Blak-Ray B100A lamp equipped with a 366-nm narrow bandpass filter, and a Rayonet Photoreactor equipped with 16 fluorescent tubes with $\lambda_{max} = 350$ nm or with 254-nm low pressure Hg lamps. Lamp intensities for quantum yield measurements were determined using ferrioxalate actinometry (32) and were of the order of 6.36×10^{-7} ein/min at 366 nm.

Spectral Measurements. IR spectra were recorded on a Perkin Elmer 621 grating IR spectrophotometer using 0.5-nm pathlength NaCl solution IR cells. These cells were sealed with serum caps and purged with nitrogen in order to record spectra of air sensitive compounds. A Varian 1400 gas chromatograph with a 25-foot SE-30 column was used for analysis of the hexene solutions. A Dupont Model 830 liquid chromatograph equipped with a 254-nm UV detector and a 100 cm $\times$ 2.1 mm Corasil II-C_{18} reverse phase column was used for analysis of the $HFeCo_3$-$(CO)_{10}(PPh_3)_2$ photolysis solutions. Mass spectra were recorded with an AEI MS902 mass spectrometer, and electronic absorption spectra were recorded on a Cary 17 spectrophotometer using 1.0-cm pathlength quartz cells. Spectra at 77°K were recorded in 5:5:2 diethyl ether–isopentane–ethanol (EPA) or 1:1 diethyl ether–isopentane solution using a quartz Dewar. Low temperature spectra were corrected for solvent contraction.

Results

Electronic Absorption Spectra. Electronic absorption spectral data for $HCCo_3(CO)_9$, $CH_3CCo(CO)_9$, $CH_3CCo_3(CO)_8PPh_3$, $HFeCo_3$-$(CO)_{12}$, and $HFeCo_3(CO)_{10}(PPh_3)_2$ are set out in Table I. The 300° and 77°K spectra of $CH_3CCo_3(CO)_9$ are shown in Figure 1, and these spectra

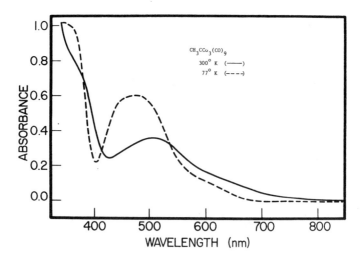

Figure 1. Electronic absorption spectra of $CH_3CCo_3(CO)_9$ at 300°K (———) and 77°K (- - -) in EPA solution

Table I. Electronic Absorption Spectral Data

Cluster	300°K[a]		77°K[b]	
	λ max (nm)	ε max	λ max (nm)	ε max
$HCCo_3(CO)_9$	505	1590	470	2240
	370 (sh)	3740	360	3780
	290 (sh)	13,500	—	—
$CH_3CCo_3(CO)_9$	510	1840	475	2430
			448	2320
	370 (sh)	3680	360	3750
	295 (sh)	13,600	—	—
$CH_3CCo_3(CO)_8PPh_3$	502[c]	2490	485	3830
			466	3770
	400 (sh)	5320	390	8080
	310 (sh)	16,170	—	—
$HFeCo_3(CO)_{12}$	528	3980	591[d]	—
			525	5150
	380 (sh)	6940	377	9510
	324 (sh)	11,000	324	13,270
	280 (sh)	12,400	—	—
$HFeCo_3(CO)_{10}(PPh_3)_2$	680 (sh)[c]	2450	660	4070
	592	4590	580	7790
	381	12,140	380	17,370
	308	16,350	305	16,290
$Na[Co_3(CO)_{10}]$	505 (sh)	--	—	—
	371 (sh)	—	—	—

[a] 2,2,4-Trimethylpentane solution.
[b] Diethyl ether–isopentane–ethyl alcohol (5:5:2) solution.
[c] CH_2Cl_2 solution.
[d] Diethyl ether–isopentane (1:1) solution.

are representative of all the compounds examined. Each of the clusters shows a broad band in the visible spectral region with a maximum between 502–592 nm and two well-defined shoulders at 370–400 nm and 290–325 nm on a rising absorption into the UV. These bands sharpen considerably upon cooling to 77°K, and the broad absorption in the visible region is seen as two distinct transitions. To aid interpretation, spectral data for $Na[Co_3(CO)_{10}]$ were obtained and are included for comparison in Table I.

Photolysis of $HCCo_3(CO)_9$. Irradiation of degassed solutions of $HCCo_3(CO)_9$ with 254-, 350-, or 366-nm light leads only to very slow decomposition. For example, a benzene solution of the cluster that had been irradiated at 350 nm for 6 hr showed only a 2% decrease in the principal absorption band at 505 nm. The cluster showed similar stability when irradiated with 254-nm light.

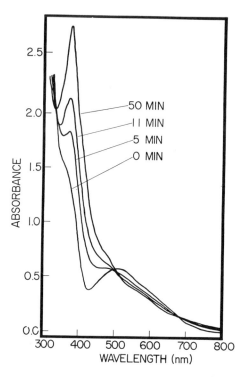

Figure 2. Electronic absorption spectral changes during 366-nm irradiation of $HCCo_3(CO)_9$ in hexane solution under a hydrogen atmosphere

The cluster is quite photosensitive, however, when irradiated with visible or UV light in the presence of hydrogen. As the irradiation proceeds, solutions rapidly change color from reddish-purple to yellow-brown, and the electronic absorption spectral changes shown in Figure 2 are obtained. The final spectrum shown is identical to that of a pure sample of $Co_4(CO)_{12}$ and indicates its formation. Further evidence for production of $Co_4(CO)_{12}$ comes from monitoring the photolysis in the IR spectral region where a smooth decrease in the ν_{CO} bands attributable to $HCCo_3(CO)_9$ is observed, and bands at 2063, 2055, 2038, 2028, and 1867 cm^{-1}, characteristic of $Co_4(CO)_{12}$ (33), grow in. Mass spectral analysis of the gases above irradiated solutions showed formation of substantial quantities of methane, and hence the overall reaction is that summarized in Reaction 9. Production of $Co_4(CO)_{12}$ was shown to be quantitative

$$HCCo_3(CO)_9 + H_2 \xrightarrow{h\nu} Co_4(CO)_{12} + CH_4 \qquad (9)$$

by measuring the final intensity of its characteristic absorption band at 373 nm ($\epsilon = 10{,}500$) ($34, 35$). The photoreaction can be induced by irradiation with $\lambda \leq 500$ nm, but spectral overlap has precluded quantum yield measurements. Importantly, no reaction was observed when solutions of $HCCo_3(CO)_9$ were stored under a hydrogen atmosphere in the dark, although decomposition to $Co_4(CO)_{12}$ plus other unidentified products did occur when hexane solutions of $HCCo_3(CO)_9$ were heated to reflux under a hydrogen purge.

Photolysis of benzene, pentane, or isooctane solutions of $HCCo_3(CO)_9$ under a 3:1 mixture of hydrogen and carbon monoxide also leads to rapid photoreaction. Mass spectral analysis confirmed the production of CH_4, but other carbon-containing products were not detected. The growth of bands at 2070, 2043, 2025, and 1858 cm^{-1} in the IR (36) and the appearance of a band at 350 nm (37) in the electronic absorption spectrum indicate that $Co_2(CO)_8$ is produced. Formation of $Co_2(CO)_8$ was shown to be quantitative by monitoring the 350 nm ($\epsilon = 5500$) (37) absorption band, and the overall reaction under H_2:CO is summarized in Reaction 10. The reaction can be induced by photolysis with $\lambda \leq 500$

$$HCCo_3(CO)_9 + H_2 + CO \xrightarrow{h\nu} CH_4 + Co_2(CO)_8 \quad (10)$$

nm, and the quantum yield of disappearance of $HCCo_3(CO)_9$ measured at 366 nm under a 3:1 H_2:CO atmosphere is 0.03. Photolysis of the cluster under a pure carbon monoxide atmosphere leads to very slow formation of $Co_2(CO)_8$. For example, after 3 hr photolysis with 350 nm, only 20% disappearance of $HCCo_3(CO)_9$ was detected whereas under an H_2:CO atmosphere, the reaction is complete within 1 hr. The organic products from photolysis under carbon monoxide were not identified.

It was of obvious importance to determine the source of the methane produced. Mass spectral analysis of the gases above a solution of $HCCo_3(CO)_9$ that had been irradiated under a D_2:CO atmosphere showed only the presence of CHD_3, and no CH_2D_2, CDH_3, or CD_4 were detected. We thus conclude that all the methane formed comes from the initial apical CH group and no hydrogenation of carbon monoxide occurs.

The conversion of $HCCo_3(CO)_9$ into $Co_4(CO)_{12}$ cannot be direct and must proceed through formation of several intermediates. This is evidenced by our observation of a slow continual 2–5% increase in intensity of the 373-nm band of $Co_4(CO)_{12}$ for several days after irradiation had ceased. Attempts to identify the initially produced fragments largely have been unsuccessful, but we have observed that at least one is a catalyst for the isomerization of 1-hexene. When $HCCo_3(CO)_9$ is irradiated in neat 1-hexene under hydrogen, isomerization to *cis*- and *trans*-2-hexene (1:2.5) occurred but with no hydrogenation. Photolysis in degassed 1-hexene gave no change in the cluster nor isomerization. Hydrogen is thus essential for the reaction. Experiments have shown that the isomerization is photocatalytic since it continued long after the irradiation had ceased. For example, in a typical experiment a sample was irradiated for 1 hr, giving 30% isomerization. After storage of the irradiated solution in the dark for four days, the isomerization had increased to 85%. In a separate experiment, we also observed that photolysis of $Co_4(CO)_{12}$ under hydrogen in 1-hexene gave isomerization but at a rate much slower than that for $HCCo_3(CO)_9$.

The photochemical properties of $HCCo_3(CO)_9$ also were examined in the presence of excess PPh_3. At room temperature, a slow thermal substitution of CO by PPh_3 occurs, giving approximately 10% conversion in 24 hr. When degassed benzene solutions containing excess PPh_3 were irradiated with 350-nm light, a very rapid reaction was indicated by the electronic absorption spectral changes. IR and TLC analysis of a solution irradiated for 45 min showed the formation of $HCCo_3(CO)_8(PPh_3)$, $HCCo_3(CO)_7(PPh_3)_2$, and at least one other unidentified product. It appears that substitution of CO by PPh_3 can be photoinduced, but that progressive photochemical reactions complicate the process. These reactions were not examined further.

Photolysis of $CH_3CCo_3(CO)_9$. The photochemical properties of $CH_3CCo_3(CO)_9$ parallel those of $HCCo_3(CO)_9$. No photoreaction was observed in the absence of hydrogen, but when $CH_3CCo_3(CO)_9$ was irradiated under a hydrogen atmosphere, nearly quantitative formation of $Co_4(CO)_{12}$ occurred. The organic products of the reaction, as identified by mass spectroscopy, were ethylene and ethane in a ratio estimated as 1:2. When the irradiation was conducted under a deuterium atmosphere, scrambling of hydrogen and deuterium was found in the organic products. Surprisingly, no reaction occurred when the complex was irradiated under a mixture of hydrogen and carbon monoxide. The quantum yield of disappearance of $CH_3CCo_3(CO)_9$ under hydrogen could not be measured. However, relative rate experiments indicate that it is at least one order of magnitude less than that of $HCCo_3(CO)_9$ under comparable conditions.

Photolysis of $HFeCo_3(CO)_{12}$ and $HFeCo_3(CO)_{10}(PPh_3)_2$. Although $HFeCo_3(CO)_{12}$ is photosensitive, its reactions are inefficient and the chemistry is not as simple as that observed for $HCCo_3(CO)_9$. Photolysis of degassed benzene or isooctane solutions with 350 nm leads to formation of $Co_4(CO)_{12}$ over a period of several days. A similar reaction is observed under a hydrogen atmosphere. When irradiated under carbon monoxide, slow but nearly quantitative production of $Co_2(CO)_8$ occurred. The iron-containing products could not be isolated in a state of purity sufficient for identification.

The photochemistry of $HFeCo_3(CO)_{10}(PPh_3)_2$ is also complex and inefficient. Photolysis of degassed benzene solutions of $HFeCo_3(CO)_{10}$-$(PPh_3)_2$ leads, over a period of several days, to cluster fragmentation and formation of $Co_2(CO)_6(PPh_3)_2$, as evidenced by monitoring the electronic absorption band of the dimer at 392 nm (37). High pressure LC analysis (38) of a degassed benzene solution that had been irradiated for two days showed the presence of $HFeCo_3(CO)_{10}(PPh_3)_2$, $Co_2(CO)_6$-$(PPh_3)_2$, and at least two other unidentified products.

Thermal Reactivity of $CH_3CCo_3(CO)_8PPh_3$ and $CH_3CCo_3(CO)_7$-$(PPh_3)_2$. The triphenylphosphine derivatives of $CH_3CCo_3(CO)_9$ were prepared to examine their photoreactivity. They proved too unstable thermally, however, to examine under our reaction conditions. Although $CH_3CCo_3(CO)_8PPh_3$ was stable indefinitely in degassed isooctane solution, it slowly decomposed when placed under a hydrogen atmosphere. $CH_3CCo_3(CO)_7(PPh_3)_2$ was even more reactive, giving slow decomposition even in degassed benzene solution. The electronic absorption spectrum taken after nearly all the original cluster had disappeared showed a band at 392 nm characteristic (37) of $Co_2(CO)_6(PPh_3)_2$ and shoulders at 318 and 333 nm. An immediate thermal reaction occurred when hydrogen was admitted to a benzene solution of $CH_3CCo_3(CO)_7(PPh_3)_2$, and the electronic absorption spectrum of this solution indicated formation of $Co_2(CO)_6(PPh_3)_2$, along with other products.

Discussion

Electronic Absorption Spectra. Examination of the available literature (19, 20, 21) leads to the observation that nearly all of the $YCCo_3$-$(CO)_9$ clusters are similar in color, purple or reddish-purple, regardless of the nature of the Y group. This is illustrated by the spectral data summarized in Table I. The spectra of $HCCo_3(CO)_9$ and $CH_3CCo_3(CO)_9$ are virtually identical. Further, the spectrum of $HFeCo_3(CO)_{12}$ is remarkably similar to that of $HCCo_3(CO)_9$, showing only a slight red shift in the lower two absorption bands. This latter observation is surprising since, in effect, the apical HC group is replaced by the vastly different $Fe(CO)_3$; yet the spectral pattern does not change. A similar spectrum also is obtained for $[Co_3(CO)_{10}]^-$, in which a triply bridging carbonyl replaces the CY group, and substitution of phosphines for carbonyls in $CH_3CCo_3(CO)_9$ and $HFeCo_3(CO)_{12}$ leads only to an energy shift in the lower bands.

The similar color of nearly all the $YCCo_3(CO)_9$ clusters and the spectral data reported herein clearly indicate that the nature of the apical group perturbs only slightly, if at all, the low-lying electronic transitions in these clusters. Hence the transitions must be localized primarily within the Co_3 framework. Furthermore, the conclusion that the lowest

Scheme 1

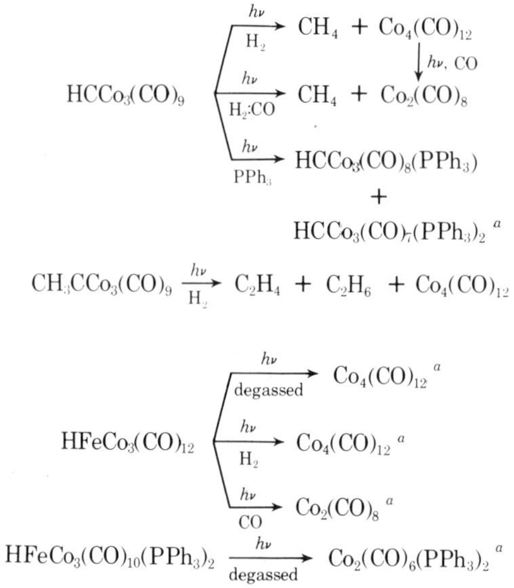

$$HCCo_3(CO)_9 \begin{cases} \xrightarrow[H_2]{h\nu} CH_4 + Co_4(CO)_{12} \quad \downarrow h\nu, CO \\ \xrightarrow[H_2;CO]{h\nu} CH_4 + Co_2(CO)_8 \\ \xrightarrow[PPh_3]{h\nu} HCCo_3(CO)_8(PPh_3) \\ \qquad\qquad + \\ \qquad HCCo_3(CO)_7(PPh_3)_2 \; ^a \end{cases}$$

$$CH_3CCo_3(CO)_9 \xrightarrow[H_2]{h\nu} C_2H_4 + C_2H_6 + Co_4(CO)_{12}$$

$$HFeCo_3(CO)_{12} \begin{cases} \xrightarrow[degassed]{h\nu} Co_4(CO)_{12} \; ^a \\ \xrightarrow[H_2]{h\nu} Co_4(CO)_{12} \; ^a \\ \xrightarrow[CO]{h\nu} Co_2(CO)_8 \; ^a \end{cases}$$

$$HFeCo_3(CO)_{10}(PPh_3)_2 \xrightarrow[degassed]{h\nu} Co_2(CO)_6(PPh_3)_2 \; ^a$$

^{a} Plus unidentified products.

unoccupied molecular orbital is localized within the Co_3 framework is supported by recent ESR studies (39, 40). The ESR spectra of the one-electron reduction products from several $YCCo_3(CO)_9$ clusters have been shown to be compatible with the added electron residing in an orbital encompassing only the three cobalt atoms, with little spin-density delocalized onto the apical carbon. A molecular orbital analysis performed by Strouse and Dahl (41) for the $XCo_3(CO)_9$ (X = S, Se) clusters would suggest that the observed bands in the $YCCo_3(CO)_9$ clusters are attributable to transitions from nonbonding to antibonding metal–metal orbitals.

Photochemistry. The photochemical reactions observed in this study are summarized in Scheme I. In general, they represent photoinduced declusterification followed by redistribution of the $Co(CO)_3$ units. In the absence of CO, $Co_4(CO)_{12}$ is obtained whereas $Co_2(CO)_8$ results when CO is available. It is further necessary to have hydrogen present to observe rapid declusterification of $HCCo_3(CO)_9$ and $CH_3CCo_3(CO)_9$ although $HFeCo_3(CO)_{12}$ and $HFeCo_3(CO)_{10}(PPh_3)_2$ are photoactive even in degassed solution.

If, as we propose, the lowest excited states in these clusters correspond to population of metal–metal antibonding orbitals, the overall bonding in the Co_3 triangle should be greatly weakened upon excitation.

It is reasonable to assume that cleavage of one of the cobalt–cobalt bonds would result to generate a diradical such as that shown in Structure 4.

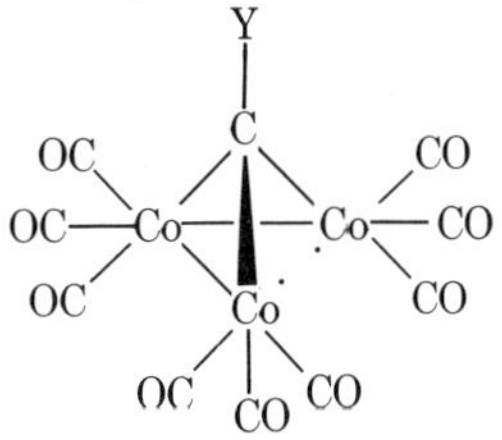

Structure 4

Support for the assumption of metal–metal bond cleavage comes from studies of the $Ru_3(CO)_{12}$ (*42*) and $H_3Re_3(CO)_{12}$ (*43, 44*) trimers, each of which gives photoinduced fragmentation to monomeric and dimeric compounds, respectively, and from the numerous studies (*45*) of metal–metal dimers, which have been demonstrated to give cleavage of the metal–metal bond upon photolysis.

Two distinctly different mechanisms, each of which begins from Structure 4, can be written to explain the net photochemical transformation. The reaction can proceed by hydrogen addition across the open cobalt–cobalt bond to generate a dihydride such as that shown in Structure 5. This complex could then eliminate hydrogen to reform $YCCo_3$-

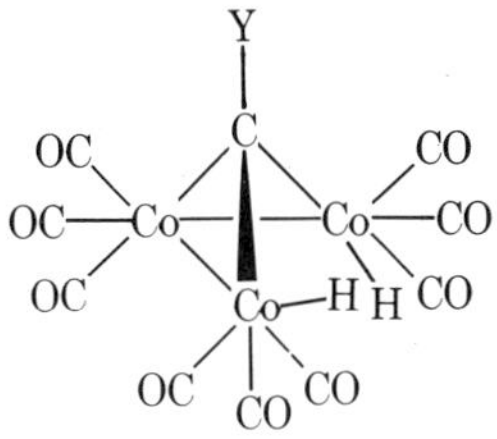

Structures 5 and 6

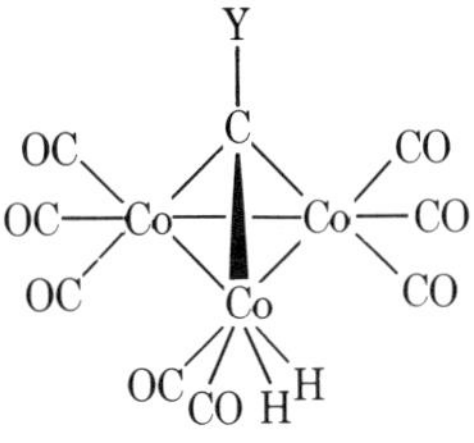

$(CO)_9$, or hydrogen could migrate to the CY group. In the presence of additional hydrogen, this migration would likely be followed by further rapid hydrogenation to form H_3CY and $HCo(CO)_3$ fragments. Combination of these cobalt fragments with concomitant hydrogen loss would generate $Co_4(CO)_{12}$ in the absence of CO, or $Co_2(CO)_8$ in the presence of CO. The proposed formation of $HCo(CO)_3$ is supported by the observation of catalytic olefin isomerization when $HCCo_3(CO)_9$ is irradiated under hydrogen in the presence of 1-hexene. $HCo(CO)_3$ has been implicated as the key intermediate in the isomerization of olefins catalyzed by $HCo(CO)_4$ (*46, 47, 48*).

The second mechanism would invoke thermal or photochemical dissociation of carbon monoxide from the radical centers that result from initial cleavage of the cobalt–cobalt bond. Such dissociation would leave open a coordination site on one of the cobalt atoms even if it were followed by rapid reformation of the metal–metal bond. Oxidative addition of hydrogen at this open site to give an intermediate such as Structure 6 could occur, followed by hydrogen transfer to the apical carbon to yield declusterification as in the mechanism discussed above. Such an overall mechanism has close parallel to that proposed by Byers and Brown (49) to account for carbon monoxide substitution in $HRe(CO)_5$. In this study, thermal dissociation of carbon monoxide from the $\cdot Re(CO)_5$ radical intermediate was suggested to occur much faster than from the non-radicals $HRe(CO)_5$ and $Re_2(CO)_{10}$.

Support for the second mechanism comes from our observation that photolysis can lead to substitution of carbon monoxide by PPh_3 in $HCCo_3$-$(CO)_9$. Furthermore, if the second mechanism is operative, then the net photoinduced declusterification should be retarded in the presence of carbon monoxide, since carbon monoxide would compete with hydrogen for the open coordination site. Although $CH_3CCo_3(CO)_9$ gives Co_4-$(CO)_{12}$ when irradiated under an hydrogen atmosphere, the declusterification was completely inhibited under a 3:1 H_2:CO atmosphere. In direct contrast, however, no carbon monoxide inhibition of the declusterification of $HCCo_3(CO)_9$ was observed, and identical rates were obtained for photolysis under 10:25:75—$CO:H_2:N_2$ and 75:25—$CO:H_2$ atmospheres. It also can be argued that phosphine substitution is not inconsistent with the first mechanism since phosphine attack directly at the open coordination site that results from metal–metal bond cleavage and subsequent loss of carbon monoxide would lead to substitution.

The available experimental evidence does not allow us to define the overall mechanism for the photoinduced declusterification. The electronic absorption spectra and precedence from previous studies of metal trimers (42, 43, 44) and dimers (45) strongly suggest that the primary photochemical reaction is cleavage of one of the cobalt–cobalt bonds. Importantly, these studies do show that it is necessary to have hydrogen present to induce declusterification of $YCCo_3(CO)_9$ since fragmentation can proceed only by removal of the CY group, and such removal is best accomplished through hydrogenation.

Acknowledgment

We thank the donors of the Petroleum Research Fund administered by the American Chemical Society and the National Science Foundation (Grant CHE 7505909) for support of this research.

Literature Cited

1. King, R. B., *Prog. Inorg. Chem.* (1972) **15**, 287.
2. Chini, P., *Inorg. Chim. Acta Rev.* (1968) **2**, 31.
3. Chini, P., Longoni, G., Albano, V. G., *Adv. Organomet. Chem.* (1976) **14**, 285.
4. Wrighton, M. S., *Chem. Rev.* (1974) **74**, 401.
5. Johnson, B. F. G., Kelland, J. W., Lewis, J., Rehani, S. K., *J. Organomet. Chem.* (1976) **113**, C42.
6. Johnson, B. F. G., Lewis, J., Twigg, M. V., *J. Chem. Soc., Dalton Trans.* (1975) 1876.
7. Kruczynski, L., Martin, J. L., Takats, J., *J. Organomet. Chem.* (1974) **80**, C9.
8. Cotton, F. A., Hunter, D. L., *J. Am. Chem. Soc.* (1976) **98**, 1413.
9. Bruce, M. I., Cooke, M., Green, M., Westlake, D. J., *J. Chem. Soc. A* (1969) 987.
10. Knox, S. A. R., Stone, F. G. A., *J. Chem. Soc. A* (1969) 2559.
11. Ibid. (1969) 3147.
12. Ibid. (1971) 2874.
13. Brookes, A., Knox, S. A. R., Stone, F. G. A., *J. Chem. Soc. A* (1971) 3469.
14. Cullen, W. R., Harbourne, D. A., Liengme, B. V., Sams, J. R., *J. Am. Chem. Soc.* (1968) **90**, 3293.
15. Cullen, W. R., Harbourne, D. A., *Inorg. Chem.* (1970) **9**, 1839.
16. Cullen, W. R., Harbourne, D. A., Liengme, B. V., Sams, J. R., *Inorg. Chem.* (1970) **9**, 702.
17. Johnson, B. F. G., Kelland, J. W., Lewis, J., Rehani, S. K., *J. Organomet. Chem.* (1976) **113**, C42.
18. Bock, C. R., Wrighton, M. S., *Inorg. Chem.* (1977) **16**, 1309.
19. Palyi, G., Piacenti, F., Marko, L., *Inorg. Chim. Acta Rev.* (1970) **4**, 109.
20. Penfold, B. R., Robinson, B. H., *Acc. Chem. Res.* (1973) **6**, 73.
21. Seyferth, D., *Adv. Organomet. Chem.* (1976) **14**, 97.
22. Sutton, P. W., Dahl, L. F., *J. Am. Chem. Soc.* (1967) **89**, 261.
23. Matheson, T. W., Robinson, B. H., Tham, W. S., *J. Chem. Soc. A* (1971) 1457.
24. Huie, B. T., Knobler, C. B., Kaesz, H. D., *J. Chem. Soc., Chem. Commun.* (1975) 684.
25. Ercoli, R., Santambrogio, E., Casagrande, G. T., *Chim. Ind. (Milan)* (1962) **44**, 1344.
26. Dent, W. T., Duncanson, L. A., Guy, R. G., Reed, H. W. B., Shaw, B. L., *Proc. Chem. Soc.* (1961) 169.
27. Chini, P., Colli, L., Peraldo, M., *Gazz. Chim. Ital.* (1960) **90**, 1005.
28. Cooke, C. G., Mays, M. J., *J. Chem. Soc., Dalton Trans.* (1975) 455.
29. Chini, P., Albano, V., Martinengo, S., *J. Organomet. Chem.* (1969) **16**, 471.
30. Fieldhouse, S. A., Freeland, B. H., Mann, C. D. M., O'Brien, R. J., *Chem. Commun.* (1970) 181.
31. Manning, A. R., *J. Chem. Soc. A* (1968) 1135.
32. Hatchard, C. G., Parker, C. A., *Proc. Roc. Soc., Ser. A* (1956) **235**, 518.
33. Bor, G., Marko, L., *Spectrochim. Acta* (1960) **16**, 1105.
34. Frazier, C., Ph.D. thesis, California Institute of Technology (1975).
35. Gray, H. B., Trogler, W. C., private communication.
36. Cable, J. W., Nyholm, R. S., Sheline, R. K., *J. Am. Chem. Soc.* (1954) **76**, 3373.
37. Abrahamson, H. B., Frazier, C. C., Ginley, D. S., Gray, H. B., Libienthal, J., Tyler, D. R., Wrighton, M. S., *Inorg. Chem.* (1977) **16**, 1554.
38. Enos, C. T., Geoffroy, G. L., Risby, T. H., *J. Chromatogr. Sci.* (1977) **15**, 83.

39. Kotz, J. C., Petersen, J. V., Reed, R. C., J. Organomet. Chem. (1976) **120**, 433.
40. Peake, B. M., Robinson, B. H., Simpson, J., Watson, D. J., Inorg. Chem. (1977) **16**, 405.
41. Strouse, C. E., Dahl, L. F., J. Am. Chem. Soc. (1971) **93**, 6032.
42. Johnson, B. F. G., Lewis, J., Twigg, M. V., J. Organomet. Chem. (1974) **67**, C75.
43. Geoffroy, G. L., Henderson, R., Gladfelter, W. L., unpublished data.
44. Kaesz, H. D., private communication.
45. Wrighton, M. S., Top. Curr. Chem. (1976) **65**, 37, and references therein.
46. Breslow, D. S., Heck, R. F., Chem. Ind. (1960) 467.
47. Heck, R. F., Breslow, D. S., J. Am. Chem. Soc. (1961) **83**, 4023.
48. Heck, R. F., J. Am. Chem. Soc. (1963) **85**, 651.
49. Byers, B. H., Brown, T. L., J. Am. Chem. Soc. (1977) **99**, 2527.

RECEIVED September 20, 1977.

Photochemistry of Bis(dinitrogen)bis[1,2-bis(diphenylphosphino)ethane]molybdenum

T. ADRIAN GEORGE, DAVID C. BUSBY, and S. D. ALLEN ISKE, JR.
University of Nebraska–Lincoln, Lincoln, NB 68588

> Studies of the reactions of alkyl bromides and iodides with a bisdinitrogen complex of molybdenum, trans-$Mo(N_2)_2$-$[(C_6H_5)_2PCH_2CH_2P(C_6H_5)_2]_2$ (1) have uncovered a fascinating new class of complexes. The 1,3-addition of the alkyl halide across a Mo–N_2 moiety produces surprisingly stable alkyldiazenido derivatives. The alkyldiazenido ligand can be protonated to form the corresponding alkylhydrazido derivative. Compound 1 undergoes $^{14}N_2/^{15}N_2$ exchange at a rate approximately 4.5 times faster in light than in total darkness. This rate differential also is observed in the formation of alkyldiazenido complexes. The electronic absorption spectra of dinitrogen, alkyldiazenido, and alkylhydrazido derivatives have been measured in solution at ambient temperature. Assignment of Mo → N_2, Mo → P, and Mo → N_2R metal-to-ligand charge transfer transitions have been made for the first two classes of compounds.

In 1974 in consecutive reports at the First International Symposium on Nitrogen Fixation, the reactions of alkyl bromides and alkyl iodides with $Mo(N_2)_2(dppe)_2$, Compound 1 [dppe = $(C_6H_5)_2PCH_2CH_2P(C_6H_5)_2$], in the presence of light were revealed (1, 2). The reaction of methyl iodide with 1 (Reaction 1) in benzene solution was reported (1) to

$$Mo(N_2)_2(dppe)_2 + CH_3I \rightarrow MoI(N_2CH_3)(dppe)_2 + N_2 \quad (1)$$

occur in both the light and the dark, but the dark reaction occurred at a slower rate. The major product of these reactions, namely the novel alkyldiazenido complex of molybdenum, has been characterized by x-ray diffraction studies for $MoI(N_2C_6H_{11})(dppe)_2$ (3, 5) and $MoI(N_2C_8H_{17})$-$(dppe)_2$ (5). The alkyldiazenido ligand is bonded in a singly bent fashion to molybdenum, trans to the iodide ligand. The structure of the alkylhydrazido derivative, [MoI($N_2HC_8H_{17}$)(dppe)$_2$]I, resulting from the protonation of the carbon-bound nitrogen also has been reported (4). The solid state structure of the parent dinitrogen complex, 1, has been determined (6) and shows a trans configuration of the two dinitrogen ligands. IR (7, 8) and phosphorus-31 NMR spectral (8) studies show that this configuration is retained in solution. However, a weak band attributed to the symmetry-forbidden A_{1g} $\nu(N_2)$ mode is observed in the IR spectrum and suggests some deviation from D_{4h} symmetry.

The first report of a specific photochemical reaction of 1 occurred in 1972 (9). A tetrahydrofuran solution of 1 in a borosilicate glass vessel at 20°C together with carbon monoxide were irradiated with a Hanovia 550-W mercury lamp. Approximately 50% yield of cis-Mo(CO)$_2$(dppe)$_2$ resulted after 15 min, with no trace of starting material remaining. On the other hand, without irradiation the same reaction occurs but much more slowly (10, 11, 12). Dihydrogen reacts rapidly with 1 without irradiation to replace both dinitrogen ligands to give MoH$_4$(dppe)$_2$ (13).

The majority of reactions of metal complexes containing coordinated dinitrogen result in the loss of dinitrogen as N_2. Dinitrogen in 1 is similarly replaced in many reactions (vide supra). The first report of the spontaneous loss of dinitrogen from 1 was in 1973 (8). When argon was bubbled through a benzene solution of 1, all dinitrogen was eventually lost. This result suggested that in solution there may be a monodinitrogen- and a nondinitrogen-containing species in equilibrium with 1 (Reactions 2 and 3) and that argon is sweeping out the dissociated nitrogen, causing the equilibrium to eventually produce a non-nitrogen-containing compound. In a closed system, a benzene solution of 1 has been irradiated

$$Mo(N_2)_2(dppe)_2 \rightleftarrows Mo(N_2)(dppe)_2 + N_2 \quad (2)$$

$$Mo(N_2)(dppe)_2 \rightleftarrows Mo(dppe)_2 + N_2 \quad (3)$$

at 366 nm with a 100-W Blak–Ray lamp at 18°C for 116 hr with > 95% recovery of 1 afterwards (14).

Recently, three important pieces of relevant data appeared in the literature. First, it was reported that benzonitrile reacts with 1 (Reaction 4) with a first-order rate constant of ca. 1.5×10^{-4} sec^{-1} (15). In the second case (16), the reaction of ethyl bromide with 1 in tetrahydrofuran

$$Mo(N_2)_2(dppe)_2 + C_6H_5CN \rightarrow Mo(N_2)(C_6H_6CN)(dppe)_2 + N_2 \quad (4)$$

was shown to be pseudo-first-order in 1 ($k \sim 10^{-4}$ sec^{-1}). Thirdly, $^{14}N_2$/$^{15}N_2$ exchange for 1 in tetrahydrofuran proceeds rapidly in diffuse daylight. The mechanism proposed for the ethyl bromide reaction that was consistent with all the data is shown in Reactions 5 and 6 (*16*).

$$\text{Mo}(N_2)_2(\text{dppe})_2 \rightarrow \text{Mo}(N_2)(\text{dppe})_2 + N_2 \qquad (5)$$

$$\text{Mo}(N_2)(\text{dppe})_2 + C_2H_5Br \rightarrow \text{MoBr}(N_2C_2H_5)(\text{dppe})_2 \qquad (6)$$

Introduction to the Compounds of Interest

Alkyl bromides and iodides react with 1 in benzene to give the corresponding alkyldiazenido derivatives of molybdenum (Reaction 7).

$$\text{Mo}(N_2)_2(\text{dppe})_2 + RX \rightarrow \text{MoX}(N_2R)(\text{dppe})_2 + N_2 \qquad (7)$$

$$(R = CH_3, C_4H_9, C_6H_{11}, C_8H_{17};\ X = Br, I)$$

These reactions are usually conducted under dinitrogen or in vacuo with irradiation provided by either sunlight, the laboratory fluorescent lights, or light bulbs placed around the reaction vessel. These reactions will proceed equally well, although more slowly, in total darkness. Functionally substituted alkyl halides such as ethyl haloacetates react similarly (Reaction 8) (*17*). However, ethyl 4-chlorobutyrate does not react with

$$\text{Mo}(N_2)_2(\text{dppe})_2 + XCH_2COOC_2H_5 \rightarrow$$
$$\text{MoX}(N_2CH_2COOC_2H_5)(\text{dppe})_2 + N_2 \qquad (8)$$
$$(X = Cl, Br, I)$$

1 to form an alkyldiazenido complex.

The alkyldiazenido ligand is readily protonated at the carbon-bound nitrogen atom to give the corresponding alkylhydrazido derivative (Reaction 9) (*4, 18*).

$$\text{MoI}(N_2C_8H_{17})(\text{dppe})_2 + HX \rightarrow [\text{MoI}(N_2HC_8H_{17})(\text{dppe})_2]X \qquad (9)$$
$$(X = Cl, Br, I, BF_4, PF_6)$$

The halide ion trans to the alkyldiazenido ligand is labile and can be replaced with other ligands such as another halide, azide, hydroxide, and thiocyanate ion (*14, 19, 20*).

Photochemical Aspects of the Preparation of Alkyldiazenido Complexes

Although alkyl bromides and iodides react with 1 to form alkyldiazenido complexes in the dark as well as in the light, there is a noticeable difference in the reaction rate. We looked at the $^{14}N_2/^{15}N_2$ exchange rate in the dark and compared it with exchange under our normal reaction condition (but with absence of alkyl halide) and found a significant difference. In the dark $k \sim 2.6 \times 10^{-5}$ sec^{-1} ($t_{1/2} = 7.5$ hr). Under reaction–irradiation conditions (three 150-W light bulbs) $k \sim 1.2 \times 10^{-4}$ sec^{-1} ($t_{1/2} = 1.6$ hr). Quantum yields have not been determined. A similar difference in rate of reaction has been observed in the dark vs. light reaction of n-butyl bromide with 1. The alkyldiazenido complexes formed from the nonactivated alkyl halides do not undergo decomposition once formed in the light reactions. However, increasing the temperature does promote the formation of the dihalide, e.g., $MoX_2(dppe)_2$, reducing the yield of the alkyldiazenido complex.

Electronic Absorption Spectra of Dinitrogen, Alkyldiazenido, and Alkylhydrazido Derivatives of Molybdenum

Table I contains the pertinent electronic absorption spectral data for some of the compounds of interest. Compound 1 and $Mo(N_2)_2(depe)_2$, 2 [depe $= (C_2H_5)_2PCH_2CH_2P(C_2H_5)_2$], are orange. The alkyldiazenido complexes are yellow-orange and the alkylhydrazido complexes brown.

The absorption spectra of 1 and 2 in benzene and in heptane solution, respectively, are shown in Figures 1 and 2. Compound 1 shows a broad band at 377 nm ($\epsilon = 13,900$) that tails off into the visible with a shoulder

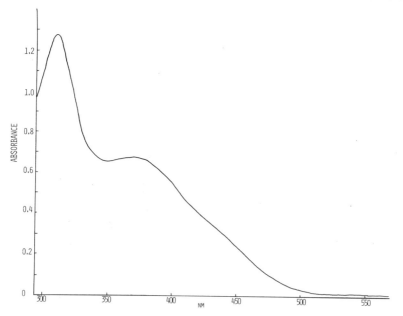

Figure 1. *Electronic absorption spectrum of $Mo(N_2)_2(dppe)_2$ in benzene solution*

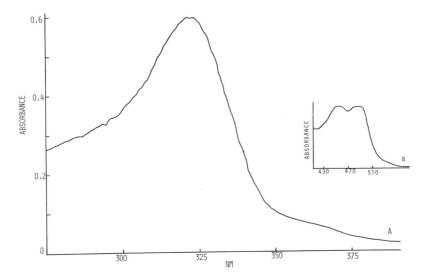

Figure 2. Electronic absorption spectrum of $Mo(N_2)_2(depe)_2$ in heptane solution

at ca. 440 nm ($\epsilon = 7,100$) and a second band at 310 nm ($\epsilon = 26,600$). The dppe ligand exhibits very intense absorptions below 300 nm because of intraligand transitions. These transitions are eliminated by replacing dppe with the bidentate tertiary alkylphosphine depe. The increased solubility of 2 in hydrocarbon solvents enabled the spectrum to be run in heptane solution. There is one intense band at 328 nm ($\epsilon = 33,000$) that tails off into the UV and the visible regions. On the visible side, there is a distinct shoulder at ca. 363 nm ($\epsilon = 2,840$). The blue shift that appears to have occurred for the 377-nm band of 1 upon going to 2 has revealed two low intensity bands at 455 and 486 nm, both with extinction coefficients of 590 M^{-1} cm^2.

In D_{4h} symmetry, the Mo(O) complexes 1 and 2 have vacant d orbitals $d_{x^2-y^2}$ (b_{1g}) and d_{z^2} (a_{1g}) together with vacant orbitals on the phosphorus atoms and dinitrogen ligands. These are shown below, assuming that the ligand field strength of the dinitrogen is greater than that of phosphine (21).

$$
\begin{array}{ll}
\text{── ──} & e_u(N_2\pi^*) \\
\text{──} & a_{2u}(Pd) \\
\text{──} & d_{x^2-y^2}(b_{1g}) \\
\text{──} & d_{z^2}(a_{1g}) \\
\text{── ──} & d_{xz}, d_{yz}(e_g) \\
\text{──} & d_{xy}(b_{2g})
\end{array}
$$

In the ground state, the b_{2g} and e_g orbitals which are π-bonding with respect to nitrogen and phosphorus are filled, $A_{1g} (b_{2g})^2 (e_g)^4$. The d–d transitions will all have the effect of weakening the Mo–N_2 and Mo–P π-bonds, and weakening the Mo–N and Mo–P σ-bonds by populating the a_{1g} and b_{1g} orbitals that are σ^* in character. We assign the two lowest energy bands of 2 to d–d transitions. If the e_g and b_{2g} orbitals have similar energies, then the two lowest transitions of 2 may actually comprise two pairs of transitions: $e_g \to a_{1g}$, $b_{2g} \to a_{1g}$ and $e_g \to b_{1g}$, $b_{2g} \to b_{1g}$. Alternatively, if the orbitals increase in energy, $b_{2g} \simeq e_g < a_{1g} << b_{1g}$, then the two bands can be attributed to $e_g \to a_{1g}$ and $b_{2g} \to a_{1g}$. The third alternative is $e_g \to a_{1g}$ and $e_g \to b_{1g}$. These d–d transitions alone would account for the loss of dinitrogen in normal light.

Both of the metal-to-ligand charge transfer (MLCT) transitions, $e_g \to a_{2u}$ (Mo $\to$ P and $e_g \to e_u$ (Mo–N_2), are symmetry allowed. The bands at 377 and 310 nm of 1 both have large extinction coefficients and could arise from the MLCT transitions. On the other hand, both transitions can be incorporated in the broad 377-nm band which exhibits at least one shoulder at about 440 nm. Compound 2 may answer the question. The $e_g \to a_{2u}$ transition is expected to be most affected when going from dppe to the more basic depe ligands. The phosphorus d orbitals of depe are of higher energy than those of dppe. The $e_g \to e_u$ transition will be less affected. Based upon this assumption, we assign the 377-nm band of 1 principally to the $e_g \to a_{2u}$ transition which is blue shifted in 2. The transition will strengthen the Mo–P π-bond and weaken the Mo–N_2 bond. This leaves the 310-nm band to be assigned to the $e_g \to e_u$ transition. The e_u set is principally associated with the nitrogen ligands and is N–N π^* (1, 22) (one e_u orbital is shown below). Working in borosilicate glass reaction vessels and with the light sources that we

$$\overset{\oplus}{N}-\overset{\ominus}{N}-Mo-\overset{\ominus}{N}-\overset{\oplus}{N}$$
$$\overset{\ominus}{}\overset{\oplus}{}\overset{\oplus}{}\overset{\ominus}{}$$

have used, these d–d and CT transitions will arise, all being Mo–N_2 bond weakening in one way or another (1, 22).

Under reaction conditions, very high yields of alkyldiazenido derivatives are achieved. Assuming that nitrogen does not insert into a preformed molybdenum–alkyl bond, one dinitrogen ligand is being retained after the other has dissociated. The solution IR spectrum suggests deviation from D_{4h} symmetry. This would have the effect of removing the degeneracy of the e_g (and e_u) orbitals. Hence promotion of an electron from what was an e_g orbital would only weaken one Mo–N_2 bond, leading to dissociation of only one dinitrogen ligand per molecule.

The electronic absorption spectral data for eight different alkyldiazenido complexes are listed in Table I. They all show a band of

Table I. Electronic Absorption Spectral Data

Compound[a]	$nm\,(\epsilon_{max})$
$Mo(N_2)_2(dppe)_2$	310 (26,600), 377 (13,900), 440 (7,100)
$Mo(N_2)_2(depe)_2$[b]	328 (33,000), 363 (2,840), 455 (590), 486 (590)
$MoBr(N_2CH_3)(dppe)_2$	364 (6,200)
$MoI(N_2CH_3)(dppe)_2$	357 (6,240)
$MoBr(N_2C_4H_9)(dppe)_2$	362 (7,160), 510 (170)
$MoI(N_2C_4H_9)(dppe)_2$	353 (6,610), 500 (160)
$Mo(N_3)(N_2C_4H_9)(dppe)_2$	358 (7,060), 515 (160)
$MoI(N_2C_6H_{11})(dppe)_2$	364 (6,520), 506 (140)
$MoOH(N_2C_6H_{11})(dppe)_2$	363 (5,670)
$MoBr(N_2CH_2COOC_2H_5)(dppe)_2$	343 (5,240), 510 (170)
$[MoI(N_2HC_6H_{11})(dppe)_2]BF_4$	316 (11,130)
$[MoI(N_2HC_8H_{17})(dppe)_2]BF_4$[c]	295 (12,800), 447 (580)
$[MoBr(N_2HCH_2COOC_2H_5)(dppe)_2]BF_4$[c]	294 (9,060), 399 (440), 550 (40)

[a] Benzene solution.
[b] Heptane solution.
[c] Tetrahydrofuran solution.

similar intensity at about 360 nm, and most of them reveal a low intensity shoulder at about 510 nm. The spectrum of $MoBr(N_2C_4H_9)(dppe)_2$ is shown in Figure 3.

The structural data in Table II show some interesting trends. Upon conversion of 1 into an alkyldiazenido compound, the Mo–P bonds

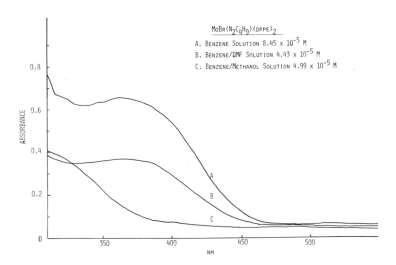

Figure 3. Electronic absorption spectrum of $MoBr(N_2C_4H_9)(dppe)_2$

Table II.

Compound	N–N	Mo–N
MoI($N_2C_6H_{11}$)(dppe)$_2$ (3,5)	1.182(19)	1.831(12)
MoI($N_2C_8H_{17}$)(dppe)$_2$ (5)	1.146(13)	1.850(12)
[MoI($N_2HC_8H_{17}$)(dppe)$_2$]I (4)	1.259(8)	1.801(5)
Mo(N_2)$_2$(dppe)$_2$ (5)	1.118(8)	2.014(5)

^{a} Bond lengths, Å. The first number in parentheses in the root mean square estimated standard deviation of an individual datum. The second and third numbers, when given, are the average and maximum deviations from the average value, respectively.

increase in length, and the remaining Mo–N bond length becomes considerably shorter. The former effect will cause the Mo–P $d\pi$–$d\pi$ overlap to be poorer while the latter effect will significantly increase the Mo–N $d\pi$–$p\pi$ interaction. X-ray photoelectron studies show that the electron density on molybdenum is essentially the same in 1 as it is in the alkyldiazenido–iodide derivatives (24).

DuBois and Hoffman (22) have developed qualitative molecular orbital diagrams which show that bending the N–N–C bond from 180° to 120° causes considerable admixing of the alkyldiazenido orbitals, resulting in the singly bent alkyldiazenido ligand being both a good σ donor and a good π acceptor. This is borne out in the solid state structures (see Table II). The in-plane N$_2$R π^* orbital is affected most strongly by the bending and decreases in energy (22). It is proposed that the band at ca. 360 nm in the spectra of the alkyldiazenido complexes is attributed to a MLCT transition; Mo(d) → N$_2$R(π^*) (vide infra).

The effect of changing the solvent on the spectra of alkyldiazenido derivatives is clearly shown in Figure 3. Figure 3 shows the absorption spectrum of MoBr(N$_2$C$_4$H$_9$)(dppe)$_2$ in benzene, benzene:DMF (1:1), and benzene:methanol (1:1) solutions. There is no difference between the first two spectra (both at 362 nm). However, the third spectrum shows considerable blue shift. Clearly, methanol is interacting significantly with part of the molecule. We believe that methanol "hydrogen bonds" to the carbon-bound nitrogen to form a pseudoalkylhydrazido derivative, -N$_2$R(HOCH$_3$). It is possible to isolate pure methanol adducts of almost all the alkyldiazenido complexes that we have prepared. The ν_{NN} stretch that appears between 1550 and 1500 cm^{-1} for the alkyldiazenido derivatives is shifted to lower energy in the methanol adducts. It has not been possible to assign the new ν_{NN}, but it must have shifted at least 50 to 100 cm^{-1} (25). In fact, the IR spectra of the methanol adducts are very similar to the alkylhydrazido complexes. An investigation of MoI(N$_2$CH$_2$COOC$_2$H$_5$)(dppe)$_2$ in methanol–, ethanol–, and 2-propanol–benzene solutions shows the extent of change increasing in

Structural Data[a]

N–C	Mo–I	Mo–P	NNC Angle
1.512(34)	2.882(2)	2.513(4,4,10)	130(1)
1.443(16)	2.882(1)	2.514(3,8,16)	128(1)
1.504(11)	2.819(1)	2.542(2,9,19)	120(1)
—	—	2.454(1)	—

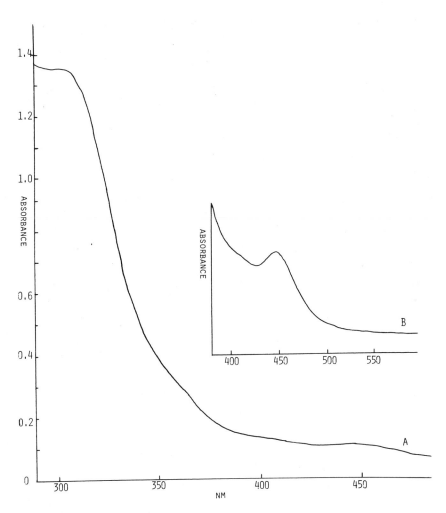

Figure 4. Electronic absorption spectrum of $[MoI(N_2HC_8H_{17})(dppe)_2]BF_4$

the order of increasing acidity of the alcohols, 2-propanol < ethanol < methanol (25).

The absorption spectra of the alkylhydrazido derivatives reported in Table I are similar to the alkyldiazenido complexes in benzene–methanol solution. Pertinent bond lengths of the solid state structure of [MoI(N_2-HC_8H_{17})(dppe)$_2$]I are given in Table II. It is interesting to compare the octyldiazenido and octylhydrazido complexes. Upon protonation, the Mo–I and Mo–N distances shorten significantly while the Mo–P and N–N bond lengths increase significantly. This will lead to stronger Mo–I and Mo–N interactions. The electronic absorption spectrum of [MoI(N_2-HC_8H_{17})(dppe)$_2$]BF_4 is shown in Figure 4.

Protonation of the carbon-bound nitrogen atom would have the effect of reorganizing the in-plane π orbitals, with only the out-of-plane orbital available for π bonding with the metal-bound nitrogen atom. This leaves the in-plane p orbital of the metal-bound nitrogen to interact exclusively with the metal to form a strong $d\pi$–$p\pi$ bond (22) (see Table II). The IR and electronic spectral data suggest that methanol is causing a similar orbital reorganization.

Acknowledgment

We would like to thank the University of Nebraska Research Council for their initial support of this work. This research also was supported in part by funds from NIH Biomedical Science Support Grant RR-07055-10 and from the National Science Foundation Grant 76-80878. Acknowledgment is made to the Donors of the Petroleum Research Fund, administered by the American Chemical Society, for partial support of this research. Thanks are also expressed to 3-M Co. and Conoco for Summer Fellowships, and to Molybdenum Climax Company for generous gifts of chemicals. We also wish to thank R. Hoffman for providing us with a copy of his paper (22) prior to publication.

Literature Cited

1. George, T. A., Iske, S. D. A., "Proceedings of the 1st International Symposium on Nitrogen Fixation," W. E. Newton, N. J. Nyman, Eds., p. 27, Washington State University, 1976.
2. Chatt, J., Diamantis, A. A., Heath, G. A., Leigh, G. J., Richards, R. L., "Proceedings of the 1st International Symposium on Nitrogen Fixation," W. E. Newton, N. J. Nyman, Eds., p. 17, Washington State University, 1976.
3. Day, V. W., George, T. A., Iske, S. D. A., *J. Am. Chem. Soc.* (1975) 97, 4127.
4. Day, V. W., George, T. A., Iske, S. D. A., Wagner, S. D., *J. Organomet. Chem.* (1976) 112, C55.
5. Day, V. W., George, T. A., unpublished data.

6. Uchida, T., Uchida, Y., Hidai, M., Kodama, T., *Acta Crystallogr.* (1975) **B31**, 1197.
7. Hidai, M., Tominari, K., Uchida, Y., Misona, A., *J. Chem. Soc., Chem. Commun.* (1969) 1392.
8. George, T. A., Seibold, C. D., *Inorg. Chem.* (1973) **12**, 2544.
9. Darensbourg, D. J., *Inorg. Nucl. Chem. Lett.* (1972) **8**, 529.
10. Hidai, M., Tominari, K., Uchida, Y., *J. Am. Chem. Soc.* (1972) **94**, 110.
11. George, T. A., Seibold, C. D., *Inorg. Chem.* (1973) **12**, 2548.
12. Aresta, M., Sacco, A., *Gazz. Chim. Ital.* (1972) **102**, 755.
13. Archer, L. J., George, T. A., *J. Organomet. Chem.* (1973) **54**, C25, and unpublished data.
14. Iske, S. D. A., Ph.D. thesis, University of Nebraska, 1977.
15. Reported in Ref. *16;* Gray, H. B., personal communication.
16. Chatt, J., Head, R. A., Leigh, G. J., Pickett, C. J., *J. Chem. Soc., Chem. Commun.* (1977) 300.
17. Busby, D. C., George, T. A., *J. Organomet. Chem.* (1976) **118**, C16.
18. Chatt, J., Diamantis, A. A., Heath, G. A., Hooper, N. E., Leigh, G. J., *J. Chem. Soc., Dalton Trans.* (1977) 688, and references cited therein.
19. Wagner, S. D., M.S. thesis, University of Nebraska, 1975.
20. Chang, M., George, T. A., unpublished data.
21. Wrighton, M., Gray, H. G., Hammond, G. S., *Mol. Photochem.* (1973) **5**, 165.
22. DuBois, D. L., Hoffmann, R., *Nouv. J. Chem.*, in press.
23. Meakin, P., Guggenberger, L. J., Peet, W. G., Muetterties, E. L., Jesson, J. P., *J. Am. Chem. Soc.* (1973) **95**, 1467.
24. Brant, P., Feltham, R. D., *J. Organomet. Chem.* (1976) **120**, C53.
25. Busby, D. C., George, T. A., unpublished data.

RECEIVED September 20, 1977.

10

Use of Transition Metal Compounds to Sensitize an Energy Storage Reaction

CHARLES KUTAL

Department of Chemistry, University of Georgia, Athens, GA 30602

A solar energy storage system based upon the interconversion of norbornadiene and quadricyclene possesses several attractive features, including high specific energy storage capacity, kinetic stability of the energy-rich photoproduct in the absence of suitable catalysts, and relatively inexpensive reactants. An inherent difficulty with the system is the lack of absorption by norbornadiene in the wavelength region of available solar radiation. Attempts to overcome this shortcoming have focused upon the use of transition metal compounds to sensitize the desired energy storage step. Criteria for suitable sensitizers are suggested, and the results of photochemical studies on a variety of systems are discussed. Evidence for the operation of two fundamentally different sensitization mechanisms also is presented.

The realization that our reserves of nonrenewable fossil fuels are dwindling rapidly has prompted the search for alternative sources of energy (1). Thus, considerable effort is currently being expended to harness and efficiently use the enormous amount of energy directly available from the sun. Because of the diffuseness, seasonal variation, and intermittent availability of sunlight, however, some means of energy storage must be coupled to the collector system to accommodate periods of peak demand. Presently, the inability to economically store solar energy for longer than a few days detracts from its otherwise attractive features. A promising solution to this problem involves the use of sunlight-induced photochemical reactions to generate storable products of high energy content that are reconvertible at will to the original material.

A hypothetical photochemical energy storage cycle is depicted in Figure 1 (2). The absorption of a photon of energy E_{ex} excites the reactant R to the photoreactive state R^*, which subsequently undergoes conversion to product P. If the overall enthalpy change associated with this process is $\triangle H$, then $\triangle H/E_{ex}$ represents that fraction of the incident photon energy which is converted to and stored as the increased chemical potential energy of P. Reversion of the system to the initial material occurs with the release of this stored energy, generally in the form of heat. The reverse step should be negligibly slow at ambient temperature because of the presence of a kinetic barrier, $\triangle H^{\ddagger}$, which must be surmounted in reaching the transition state, T. This feature ensures that energy can be stored indefinitely and is released only at elevated temperature or upon the addition of an appropriate catalyst.

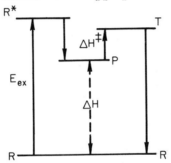

ENERGY STORAGE : $R \xrightarrow{h\nu} P$

ENERGY RELEASE : $P \xrightarrow[\text{catalyst}]{\triangle} R + \text{heat}$

Figure 1. Hypothetical photochemical energy storage cycle: R, reactant; E_{ex}, energy of incident photon; R^*, photoreactive excited state; P, photoproduct; T, transition state for thermal process, $\Delta H^{\ddagger}$, activation enthalpy; ΔH, overall enthalpy change for R–P interconversion

The net effect of cycling the energy storage–energy release steps is to convert sunlight to a more controllable and thus usable form without consuming any nonrenewable resources. Ideally the recyclable storage medium should possess the following characteristics (3):

(i) significant absorption of incident solar radiation,
(ii) high quantum efficiency of the photochemical energy storage step,
(iii) high specific energy storage capacity (heat stored per gram of photoproduct formed),
(iv) absence of destructive side reactions,
(v) ease of handling (can be cycled readily in a storage device),

(vi) synthesis from readily available and inexpensive starting materials.

Because of these stringent requirements, the number of currently known photochemical reactions which possess any promise for use in a cyclical energy storage system is understandably small. Among the most attractive candidates is the norbornadiene (NBD)-quadricyclene (Q) interconversion (Reaction 1). Both compounds are liquids with

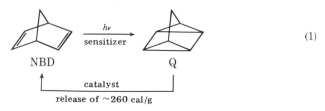

(1)

boiling points and densities similar to those of water. Although NBD itself does not absorb light in the wavelength region of available solar radiation (>300 nm), the photoreaction does occur in the presence of an appropriate spectral sensitizer with an overall efficiency of Q production approaching 100% in optimal cases (vida infra). The photoproduct, while containing some 260 cal/g excess energy over NBD (4), is stable toward thermal reversal because of orbital symmetry constraints. Exposure to certain transition metal catalysts, however, allows for the clean and rapid conversion of Q to NBD with the release of the excess energy (5). NBD is an attractive storage medium from a cost standpoint since it is prepared from commonly available chemicals (acetylene and cyclopentadiene).

The characteristics of a NBD-Q-based energy storage system recommend its use as a source of low-grade ($\sim 100°$C) heat. Some readily apparent applications along this line are the heating, cooling, and hot water production in buildings. Roughly 20% of all energy (primarily from fossil fuels) currently consumed in the United States is used for these purposes.

The Solar Energy Storage Program at the University of Georgia has been directed toward the evaluation and development of this promising system. Several related research areas are currently under investigation:

(A) *Sensitizer Development.* At the outset of the project, there was limited information available on the sensitized conversion of NBD to Q. The sensitizers that had been reported were generally weak absorbers of solar radiation, inefficient, or prone to decomposition. Thus, our efforts have concentrated on: (1) surveying a wide variety of potential candidates and, (2) studying the mechanism of sensitization to intelligently design new candidates offering improved characteristics.

(B) *Catalyst Development.* While several catalysts for the reverse reaction are known, it would be advantageous to have catalysts that are relatively inexpensive and that would offer a range of catalytic activity to select the optimal rate for the reconversion at ambient temperature. Therefore, we have been searching for new structures of high catalytic activity, high product specificity, and low cost.

(C) *Polymer Anchoring of Sensitizers and Catalysts.* The need to physically constrain the catalyst for the heat-releasing reaction to the catalytic chamber is obvious. Similar confinement of the photosensitizer to the irradiation chamber reduces the required amount of this component. Polymer immobilization also precludes undesirable interactions between the catalyst and sensitizer and facilitates their replacement in an actual device.

In the remainder of this article, the discussion will center upon the use of transition metal compounds to sensitize the photochemical production of Q. Particular emphasis is placed upon identifying those characteristics of a compound that enhance its sensitization efficiency. While any mechanistic generalizations are necessarily referenced to the NBD–Q system, they may ultimately prove to have wider applicability to analogous transformations of other olefins.

Characteristics of the Sensitized Interconversion of NBD and Q

The absorption spectrum of NBD displayed in Figure 2 is in general accord with several molecular orbital calculations (6, 7, 8) which place the first symmetry-allowed singlet transition in the molecule below 225 nm. Consequently, only relatively high energy photons are able to effect the direct photoconversion of NBD to Q, a process first reported by Dauben and Cargill in 1961 (9). Shortly thereafter Hammond and co-workers demonstrated that the same transformation can be sensitized by organic carbonyl compounds known to be effective triplet sensitizers (Table I) (10, 11). The postulated mechanism involves electronic energy transfer from the triplet state of the sensitizer to the lowest triplet of NBD. One inference which emerges from the marked drop in sensitization efficiency as the triplet energy, E_T, of the donor decreases is that the NBD triplet lies in the vicinity of 68 kcal. A calculated triplet energy of ~ 70 kcal lends support to this premise (8).

The small but measurable sensitized reversion of Q to NBD in the presence of benzophenone (Table I) and other, lower energy triplet sensitizers (11, 12) is not explicable in terms of a classical energy transfer mechanism, which would be highly endothermic and thus inefficient for these systems. Such behavior was originally attributed to "nonvertical excitation" of Q to directly form a triplet having a geometry distorted from that of the ground state molecule. Later, quenching and photochemical studies (13, 14) have provided a more detailed explanation in terms of exciplex formation between Q and the triplet sensitizer. This

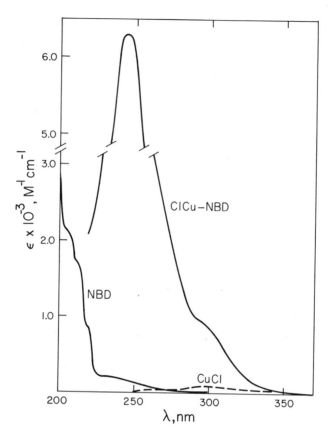

Figure 2. Spectral evidence for complex formation between CuCl and NBD in ethanol (28)

Table I. Quantum Yields for the NBD–Q Interconversion in the Presence of Triplet Sensitizers (11)

Starting Material[a]	Sensitizer	$E_T(kcal)$[b]	Quantum Yield (at 313 nm)
NBD	acetophenone	74	0.91
NBD	benzophenone	69	0.51
NBD	2-naphthaldehyde	60	0.06
NBD	2-acetonaphthone	59	0.05
Q	benzophenone	69	0.06
Q	2-naphthaldehyde	60	small

[a] NBD or Q concentrations between 0.53 and 0.64M were used.
[b] Energy of lowest-lying triplet state.

intermediate species can subsequently partition between relaxation pathways that lead to NBD or Q, or it can form biradical coupling products.

The use of aromatic hydrocarbon sensitizers with relatively long-lived singlet states (e.g., naphthalene, pyrene, 9,10-dichloroanthracene) also can effect the conversion of Q to NBD (*11*). Sensitization is compatible kinetically with the accompanying quenching of the sensitizer fluorescence, indicating that both processes involve the lowest excited singlet state in the aromatic hydrocarbon. The conclusion was drawn that the formation of an exciplex between this excited state and Q provides a pathway for sensitization. The stabilization of the exciplex by charge transfer interaction is suggested by a correlation of the rate constants for fluorescence quenching by Q and the electron affinities of the hydrocarbon sensitizers (*15*).

Until recently only two qualitative reports (*16, 17*) of inorganic compounds functioning as sensitizers for the NBD to Q conversion have appeared. Both studies concern the use of Cu(I) salts, whose known ability to strongly bind olefins suggests that ground state complexation may play a key role in the mechanism of sensitization. Complexation undoubtedly perturbs the electronic states primarily localized on NBD as well as generating new states of the metal–olefin charge transfer type. The net result of this altered distribution of electron density can be a shift of the absorption spectrum of the Cu–NBD system to longer wavelengths. Minimally, then, the complex is expected to facilitate the absorption of light by the otherwise weakly absorbing olefin. In addition, there is the possibility that the complex may provide a new, low energy pathway for the production of Q.

The mechanistic observations noted above, along with the practical constraints imposed by the demand for efficient solar energy utilization, provide a basis for formulating the following general criteria for transition metal sensitizers:

(1) significant absorption, either directly or via complex formation with NBD, in the wavelength region of available solar energy,

(2) high efficiency and specificity in sensitizing the NBD to Q energy storage step without concomitantly effecting the undesired reversion reaction,

(3) low thermal and photochemical reactivity.

In succeeding sections we describe several transition metal compounds which exhibit at least some of these desirable features.

Sensitization by Copper(I) Compounds

A large and diverse group of Cu(I) compounds has been prepared and characterized (*18, 19*). Stabilization of the $+1$ state toward oxidation and disproportionation is achieved generally through the use of various complexing and precipitating agents. Thus the soluble $CuCl_2^-$, $Cu(NH_3)_2^+$, and $Cu(CN)_3^{-2}$ complexes are stable in aqueous solution, and difficulty soluble salts such as CuCl, CuBr, CuI, and Cu_2S exist in the solid state or in contact with water. Ligands possessing low lying, vacant π^* orbitals—1,10-phenanthroline(phen), 2,2'-bipyridine (bipy), phosphines, olefins—also tend to stabilize the +1 state. While discrete two- and three-coordinate Cu(I) compounds are known, the predominant coordination number of the metal is four.

Considerable attention has been afforded Cu(I)-olefin complexes. The generally accepted Chatt–Dewar description (20) of the bonding in such complexes identifies two major components: (1) σ donation from a filled olefin π orbital to an empty metal orbital, and, (2) π back donation from a filled metal d orbital to an empty π^* olefin orbital. Both components appear to contribute to the stability of the Cu(I)–olefin bond (21). This tendency of Cu(I) to interact strongly with C–C double bonds has been exploited in photoassisted transformations of olefin molecules. Recent examples of such processes involving copper–olefin complexes include isomerization (22, 23), cycloaddition (16, 24, 25), and molecular fragmentation (26).

At the outset of our search for effective inorganic sensitizers, we screened a wide range of simple transition metal salts (27). Only those of Cu(I) afforded any appreciable conversion of NBD to Q. Direct evidence for the involvement of a Cu–NBD complex in the sensitization process emerges from spectral studies (28, 29), the main features of which are displayed in Figure 2. Neither of the parent compounds, CuCl or NBD, absorbs appreciably above ~ 230 nm, whereas a mixture of the two exhibits a new, intense band which effectively increases the absorption of the system well beyond 300 nm. This band has been assigned to a charge transfer transition within a 1:1 ClCu–NBD π complex. Complex formation ensues in a broad range of solvents (alcohols, chloroform, methylene chloride, tetrahydrofuran), but is retarded seriously by the strong solvation of Cu(I) in acetonitrile.

As shown in Table II, 313-nm irradiation of solutions containing CuCl and sufficient NBD to ensure complete complexation results in the efficient production of Q. The quantum yield exhibits no apparent dependence upon the initial NBD concentration. Likewise, there is little variation with changes in solvent composition with the notable exception of acetonitrile, where the absence of detectable conversion to Q parallels the hindrance of complex formation. Both observations are consonant with a general mechanism that features the ClCu–NBD π complex as the major photoactive species. As depicted in Scheme 1, the key role of CuCl is to complex the weakly absorbing olefin, thereby shifting its absorption spectrum into a region accessible to the irradiating light.

Table II. Quantum Yields for CuCl-Sensitized Production of Q (29)

[NBD], M[a]	Solvent	Quantum Yield (at 313 nm)
9.86 (neat)	—	0.39 ± 0.07
1.0	ethanol	0.28 ± 0.05
0.10	ethanol	0.36 ± 0.06
1.0	chloroform	0.35 ± 0.02
0.10	chloroform	0.27
0.50	tetrahydrofuran	0.42
0.50	acetonitrile	[b]

[a] [CuCl] ranged from 1 to 7 × $10^{-3}M$.
[b] No detectable quadricyclene production.

The absorption of a photon by the ClCu–NBD complex populates a charge transfer state which either directly, or following relaxation to a lower energy state or intermediate, undergoes conversion to Q. Since Q has little affinity for Cu(I), the photoactive ClCu–NBD complex is regenerated and the cycle can be repeated. Catalytic factors (mol of Q formed per mol of CuCl initially present) of at least 390 have been measured for this system (29).

Scheme 1

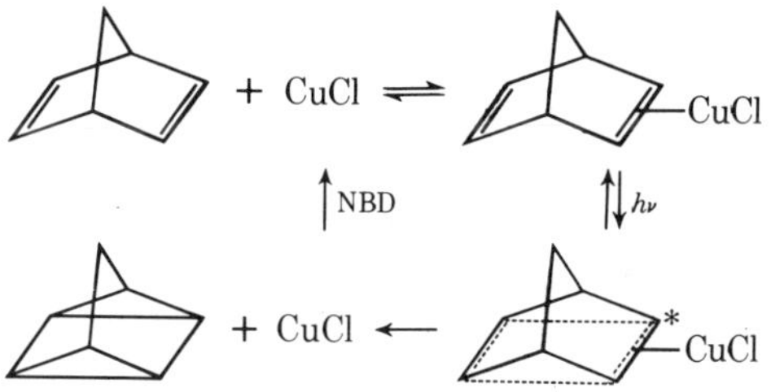

A mechanism analogous to that proposed in Scheme 1 is likely to obtain for sensitization by other simple CuX (X = Br, I, OAc) salts as well as by the more exotic compound, Cu[HB(pz)$_3$]CO (HB(pz)$_3$ is hydrotris(1-pyrazoyl)borate), Structure 1 (30, 31, 32). The latter contains a reasonably stable Cu–CO bond, presumably the result of enhanced Cu → CO backbonding in the presence of the strongly electron-donating HB(pz)$_3$ ligand. Although Cu[HB(pz)$_3$]CO exists as a monomer in solution, the CO group is substitutionally labile. Upon loss of CO, the resultant coordinatively unsaturated species can undergo dimerization or capture an added ligand, L (Reaction 2). If L = NBD, a species

$$Cu[HB(pz)_3]CO \underset{CO+}{\overset{-CO}{\rightleftharpoons}} Cu[HB(pz)_3] \underset{+L}{\overset{}{\longrightarrow}} \begin{array}{l} Cu_2[HB(pz)_3]_2 \\ Cu[HB(pz)_3]L \end{array} \quad (2)$$

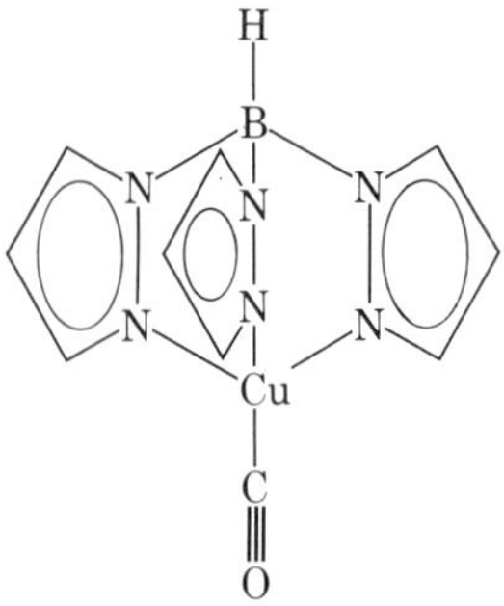

1

analogous to the proposed photoactive ClCu–NBD complex can be formed. Indeed we have observed that $Cu[HB(pz)_3]CO$ is effective (quantum yield ~ 0.2 in a $0.1M$ NBD solution in ethanol) in sensitizing Q production upon 313-nm irradiation (33).

A common drawback of the Cu(I) systems considered thus far is their susceptibility to oxidation in solution. Air must be excluded rigorously during sample preparation to avoid formation of Cu(II), which is ineffective as a sensitizer (29). Furthermore, the oxidation process may generate radicals that can initiate destructive side reactions of NBD. In an attempt to circumvent these problems, we undertook an investigation of the reasonably air stable species $Cu[P(C_6H_5)_3]_2BH_4$ and $Cu[P(C_6H_5)_2CH_3]_3BH_4$ (34). The former compound exists as Structure 2 both in the solid state and in solution, with no detectable ligand dissociation occurring in the latter environment. In contrast, $Cu[P(C_6H_5)_2CH_3]BH_4$ assumes the novel Structure 3 in the solid state (35, 36). Apparently the replacement of a C_6H_5 group by CH_3 sufficiently reduces the steric requirements of $P(C_6H_5)_2CH_3$ to permit three of the ligands to coordinate to copper. Upon dissolution in benzene, however, the IR spectrum of the compound changes noticeably and now can be interpreted in terms of bidentate coordination by BH_4^-. Formation of this second Cu–H–B bridge bond occurs at the expense of one of the originally coordinated $P(C_6H_5)_2CH_3$ ligands, since molecular weight measurements indicate that the compound is extensively dissociated.

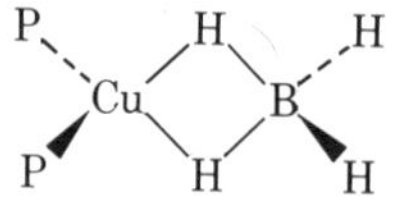

2 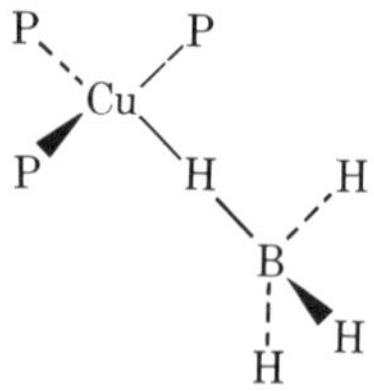   **3**

Phosphine lability in $Cu[P(C_6H_5)_3]_2BH_4$ and $Cu[P(C_6H_5)_2CH_3]_3BH_4$ was established by ^{31}P NMR spectroscopy. Thus, mixtures of each Cu(I) compound and its respective free ligand in benzene solution exhibit a single resonance whose chemical shift is dependent upon the added ligand concentration.

Both $Cu[P(C_6H_5)_3]_2BH_4$ and $Cu[P(C_6H_5)_2CH_3]_3BH_4$ sensitize the conversion of NBD to Q upon 313-nm irradiation. As depicted in Figure 3, the sensitization yields depend upon the initial NBD concentration.

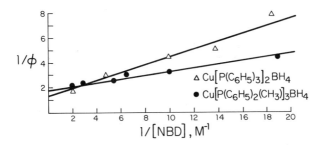

Figure 3. Reciprocal plot of sensitization efficiency (ϕ) vs. NBD concentration for $Cu[P(C_6H_5)_3]_2BH_4$ and $Cu[(C_6H_5)_2CH_3]_3BH_4$

The reciprocal of the intercept of a plot of $1/\phi$ vs. $1/[NBD]$ defines a limiting (maximum) yield of 0.76 for $Cu[P(C_6H_5)_3]_2BH_4$ and 0.56 for $Cu[P(C_6H_5)_2CH_3]_3BH_4$. In contrast, the sensitization efficiency of free $P(C_6H_5)_3$ (~ 0.01) is considerably lower.

The presence of labile phosphines in $Cu[P(C_6H_5)_3]_2BH_4$ and $Cu[P(C_6H_5)_2CH_3]_3BH_4$ is suggestive that the mechanism of sensitization involves the displacement of one or more of these ligands by NBD (Reaction 3), resulting in a complex analogous to the proposed photo-

$$Cu[P(C_6H_5)_3]_2BH_4 + NBD \rightleftharpoons BH_4[P(C_6H_5)_3]Cu\text{---}NBD + P(C_6H_5)_3 \quad (3)$$
$$\downarrow h\nu$$
$$Q$$

active species in CuX–NBD systems. Experiments designed to detect such a ligand displacement process invariably have yielded negative results, however. Thus, the quantum yield for $Cu[P(C_6H_5)_3]_2BH_4$-sensitized production of Q is unaffected in the presence of a twofold excess of $P(C_6H_5)_3$, despite the expectation that this would repress the formation of the Cu–NBD complex. Similarly, no release of coordinated phosphine is detected by ^{31}P NMR upon addition of excess NBD to solutions of $Cu[P(C_6H_5)_3]_2BH_4$ or $Cu[P(C_6H_5)_2CH_3]_3BH_4$. The additional possibility that NBD displaces one of the hydrogen atom bridges of BH_4^- also is excluded on grounds that the solution IR spectrum of $Cu[P-(C_6H_5)_2CH_3]_3BH_4$ in the presence of a considerable excess of NBD is consistent with Structure 2.

The studies cited above strongly suggest that the mechanism of sensitization by $Cu[P(C_6H_5)_3]_2BH_4$ and $Cu[P(C_6H_5)_2CH_3]_3BH_4$ is fundamentally different from that of simple CuX salts in that no direct ground state coordination of NBD to the copper atom occurs. The alternative pathway involving interaction of the photoexcited copper

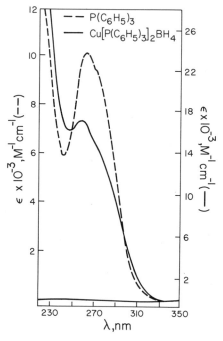

Figure 4. Electronic absorption spectra of $Cu[P(C_6H_5)_3]_2BH_4$ and $P(C_6H_5)_3$ in cyclohexane (34)

compound with NBD, presumably via electronic energy transfer or excited-state complex formation, thus merits serious consideration. While the mechanistic details of the sensitization process are currently under investigation, some interesting spectral observations deserve brief comment. Thus, Figure 4 displays the close similarity in the absorption spectra of $Cu[P(C_6H_5)_3]_2BH_4$ and $P(C_6H_5)_3$, suggesting that the 258-nm band in the copper compound be assigned as predominantly ligand localized. Likewise, both species exhibit a broad emission centered at nearly the same wavelength (~ 475 nm) in room-temperature benzene solution. These spectral data denote little energetic difference between the lowest excited states in the two compounds, in contrast to the marked disparity in their efficiencies of sensitization. We infer from such behavior that the rates of population or depopulation of the excited state which interacts with NBD must be appreciably different in the two compounds.

Our recent strategy in designing Cu(I) sensitizers has centered upon the incorporation of a strongly absorbing chromophore into the sensitizer molecule. Tetrameric [CuXPR$_3$]$_4$ clusters, for example, are cleaved by heterocyclic nitrogen bases to form colored complexes of the type CuXPR$_3$(N–N) (typically X = Cl, Br, I; R = phenyl, ethyl, n-butyl; N–N = phen or bipy) (37, 38). The characteristic colors of this class of compounds derive from the presence of a low energy Cu → N–N charge transfer transition. The intermolecular interaction of this excited state with NBD offers one possible sensitization pathway. By using bulky R substituents (e.g., n-butyl), on the other hand, it should be possible to labilize PR$_3$ toward substitution by other ligands present in solution while retaining the Cu(N–N) chromophore. This raises the intriguing possibility of producing a (N–N)XCu–NBD complex that is strongly absorbing in the visible region. More importantly, the changes in electron density attendant upon populating the Cu → N–N charge transfer state may have desirable consequences for the coordinated NBD. Thus, intramolecular electronic energy transfer from this state to the π–π* triplet of the olefin can occur, with the latter state then undergoing conversion to Q. Alternatively, the increased positive charge on copper in the charge transfer state can be delocalized partially onto the olefin. In a formal sense, the NBD acquires some carbenium ion character including, perhaps, the tendency to undergo skeletal rearrangement to Q. The sensitization properties of this intriguing class of compounds is currently under investigation.

Sensitization by Transition Metal Compounds Containing Ligands with Delocalized π Systems

Second and third row transition metal compounds containing ligands with delocalized π systems (some examples are listed in Table III) display several interesting properties. To wit—

(i) the absorption spectrum appreciably overlaps the wavelength region of available solar radiation,

(ii) the ground state is generally inert to ligand substitution processes,

(iii) the lowest lying excited state, which can be quite long-lived (~1 μsec) in room temperature fluid solution (40, 41, 42), undergoes a diverse assortment of intermolecular processes

Table III. Quantum Yields for Q Production Containing Ligands with

Compound[a]	Nature of Lowest-[b] Lying Excited State
Ru(bipy)$_3^{+2}$	$d\pi^*$
Ru(phen)$_3^{+2}$	$d\pi^*$
cis-Ir(phen)$_2$Cl$_2^+$	$d\pi^*$ or d–d
Rh(phen)$_3^{+3}$	$\pi\pi^*$
Ir(bipy)$_2$(OH$_2$)(bipy')$^{+3}$ [d]	$\pi\pi^*$
Ir(bipy)$_2$(OH)(bipy')$^{+2}$ [e]	$\pi\pi^*$

[a] [NBD] = 0.5M unless otherwise stated.
[b] Orbital designations have the following meaning: $d\pi^*$, metal-to-ligand charge transfer state; d–d, ligand field state; $\pi\pi^*$, ligand localized state. In cases where configurational mixing is important, only the predominant contributing state is listed.

with added substrates; recently reported examples include oxidation–reduction (43) and electronic energy transfer (44).

The features noted above make these transition metal compounds exceedingly attractive as potential sensitizers for the NBD to Q conversion. Thus, we have recently undertaken an exploratory study aimed at screening a variety of candidates. Our preliminary quantum yield data are collected in Table III (45) along with the orbital assignment and approximate energy, E_L, of the lowest lying excited state in each of the compounds examined.

The high sensitization efficiency of [Ir(bipy)$_2$(OH$_2$)(bipy')]$^{+3}$ (Structure 4 (42)) is surprising, at least to the extent that the energy and

orbital parentage of its lowest excited state are comparable with those of other compounds tested. It is interesting to note that electronic energy transfer from this state to the lowest triplet in NBD would be endothermic by ~ 6 kcal. While an energy barrier of this magnitude should hamper efficient sensitization because of an unfavorable Boltzmann factor (46), the long excited-state lifetime of the sensitizer would tend to ameliorate this effect. Thus, energy transfer presents a viable, but by no means exclusive, pathway for sensitization. Clearly, further work is required to resolve this question.

in the Presence of Transition Metal Compounds
Delocalized π Systems

$E_L{}^a$ (kcal)	Solvent	Quantum Yield (at 366 nm)
49	5% methanol–95% benzene	$< 2 \times 10^{-3}$
50	20% ethanol–80% cyclohexane	$< 6 \times 10^{-3}$
60	20% ethanol–80% cyclohexane	$< 4 \times 10^{-3}$
63	20% ethanol–80% cyclohexane	$< 4 \times 10^{-3}$
61	5% methanol–95% benzene	0.7
64	methanol	0.09

^{a} Approximate energy of lowest-lying excited state in compound; values from Ref. *39, 40, 41, 42*.
^{d} bipy′ denotes a monodentate-bound 2,2′-bipyridine ligand.
^{e} [NBD] = 0.1M.

Another attractive feature of $[\mathrm{Ir(bipy)_2(OH_2)(bipy')}]^{+3}$ (or any charged sensitizer) is the ease with which it can be immobilized onto a polymeric support. We recently have prepared a sulfonated resin using macroreticular (20% crosslinked) styrene–divinylbenzene copolymer beads (*47*). Immobilization is accomplished by equilibrating a solution of $[\mathrm{Ir(bipy)_2(OH_2)(bipy')}]^{+3}$ with the resin (Reaction 4). Quantitative

$$\mathrm{C_6H_4\text{–}SO_3^-K^+} + [\mathrm{Ir(bipy)_2(OH_2)(bipy')}]^{+3} \longrightarrow \mathrm{C_6H_4\text{–}SO_3^-[\mathrm{Ir(bipy)_2(OH_2)(bipy')}]^{+3}} + \mathrm{K^+} \quad (4)$$

studies concerning the effects of various parameters (such as NBD concentration, percent loading of the sensitizer on the polymer, polymer porosity, etc.) on the quantum efficiency of the polymer-bound sensitizer are just underway.

Conclusion

A diverse assortment of transition metal compounds have been shown to sensitize effectively the photochemical conversion of NBD to Q. Two fundamentally different sensitization pathways may obtain in these systems: (i) interaction of the electronically excited sensitizer with ground state NBD, (ii) photoexcitation of a ground state complex between the sensitizer and NBD. Path (i) will be favored by substitution inert, coordinatively saturated compounds possessing an excited state capable of strongly interacting with NBD. Path (ii) is more probable for compounds that form stable ground state complexes with NBD but have little tendency to interact with Q. The latter provision excludes compounds of metals which undergo facile oxidative addition (Pt(II), Pd(II), Ni(0), Rh(I) or function as good Lewis acids (Ag$^+$) since such species effectively catalyze the reversion of Q to NBD (5).

Several of the systems examined pose intriguing mechanistic questions that merit further investigation. In particular, the nature of the interaction between NBD and the excited states of sensitizers such as Cu[P(C$_6$H$_5$)$_3$]$_2$BH$_4$ and [Ir(bipy)$_2$(OH$_2$)(bipy')]$^{+3}$ requires more precise definition. It also will be of interest to test the efficacy with which some of these transition metal compounds sensitize other organic photoreactions.

In summary, we feel that considerable progress in developing transition metal sensitizers for the NBD energy storage system has been achieved. The search for new, more effective candidates as well as efforts to optimize the desirable features of existing sensitizers are continuing. Concurrent efforts to develop effective organic sensitizers (48) and reversion catalysts (49) also are being pursued.

Acknowledgment

It is my pleasure to acknowledge significant contributions from the following co-workers: P. A. Grutsch, D. P. Schwendiman, E. M. Sweet, S. C. Chang, and R. Sterling. Thanks also go to my University of Georgia colleagues, R. B. King and R. R. Hautala, for many stimulating discussions and to Richard J. Watts for kindly supplying samples of [Ir(bipy)$_2$-(OH)(bipy')]$^{+2}$ and [Ir(bipy)$_2$(OH$_2$)(bipy')]$^{+3}$. This work was supported by the National Science Foundation (MPS75-13752) and the Energy Research and Development Administration (E(38-1)-893).

Literature Cited

1. *Science* (1974) **184**, 247–386.
2. Lane, J. E., Mau, A. W.-H., Pompe, A., Sasse, W. H. F., Spurling, T. H., CSIRO Div. Appl. Org. Chem. (1977) Technical Paper No. 4.
3. Daniels, F., Heidt, L. J., Livingston, R. S., Rabinowitch, E., "Photochemistry in the Liquid and Solid States," F. Daniels, Ed., Chap. 1, John Wiley, New York, 1960.
4. Wiberg, K. B., Connon, H. A., *J. Am. Chem. Soc.* (1976) **98**, 5411.
5. Bishop, K. C., *Chem. Rev.* (1976) **76**, 461.
6. Wilcox, C. F., Winstein, S., McMillan, W. G., *J. Am. Chem. Soc.* (1960) **82**, 5450.
7. Allinger, N. L., Miller, M. A., *J. Am. Chem. Soc.* (1964) **86**, 2811.
8. Van-Catledge, F. A., *J. Am. Chem. Soc.* (1971) **93**, 4365.
9. Dauben, W. G., Cargill, R. L., *Tetrahedron* (1961) **15**, 197.
10. Hammond, G. S., Turro, N. J., Fischer, A., *J. Am. Chem. Soc.* (1961) **83**, 4674.
11. Murov, S., Hammond, G. S., *J. Phys. Chem.* (1968) **72**, 3797.

12. Hammond, G. S., Wyatt, P., Deboer, C. D., Turro, N. J., *J. Am. Chem. Soc.* (1964) **86**, 2532.
13. Gorman, A. A., Leyland, R. L., *Tetrahedron Lett.* (1972) 5245.
14. Gorman, A. A., Leyland, R. L., Rodgers, M. A. J., Smith, P. G., *Tetrahedron Lett.* (1973) 5085.
15. Solomon, B. S., Steel, S., Weller, A., *Chem. Commun.* (1969) 927.
16. Trecker, D. J., Foote, R. S., Henry, J. P., McKeon, J. E., *J. Am. Chem. Soc.* (1966) **88**, 3021.
17. Srinivasan, R., U. S. Patent **3,350,291** (1967).
18. Quin, H. W., Tsai, J. H., *Adv. Inorg. Chem. Radiochem.* (1968) **12**, 217.
19. Jardine, F. H., *Adv. Inorg. Chem. Radiochem.* (1975) **17**, 115.
20. Cotton, F. A., Wilkinson, G., "Advanced Inorganic Chemistry," 3rd ed., Chap. 23, John Wiley, New York, 1972.
21. Salomon, R. G., Kochi, J. K., *J. Am. Chem. Soc.* (1973) **95**, 1889.
22. Srinivasan, R., *J. Am. Chem. Soc.* (1964) **86**, 3318.
23. Whitesides, G. M., Goe, G. L., Cope, A. C., *J. Am. Chem. Soc.* (1969) **91**, 2608.
24. Salomon, R. G., Kochi, J. K., *J. Am. Chem. Soc.* (1974) **96**, 1137.
25. Salomon, R. G., Folting, K., Streib, W. E., Kochi, J. K., *J. Am. Chem. Soc.* (1974) **96**, 1145.
26. Salomon, R. G., Salomon, M. F., *J. Am. Chem. Soc.* (1976) **98**, 7454.
27. Kutal, C., Schwendiman, D. P., Grutsch, P. A., *Sol. Energy* (1977) **19**, 651.
28. Schwendiman, D. P., Kutal, C., *Inorg. Chem.* (1977) **16**, 719.
29. Schwendiman, D. P., Kutal, C., *J. Am. Chem. Soc.* (1977) **99**, 5677.
30. Bruce, M. I., Ostazewski, A. P. P., *Chem. Commun.* (1972) 1124.
31. Bruce, M. I., Ostazewski, A. P. P., *J. Chem. Soc., Dalton Trans.* (1973) 2433.
32. Churchill, M. R., DeBoer, B. G., Rotella, F. J., Salah, O. M. A., Bruce, M. I., *Inorg. Chem.* (1975) **14**, 2051.
33. Sterling, R., Kutal, C., unpublished data.
34. Grutsch, P. A., Kutal, C., *J. Am. Chem. Soc.* (1977) **99**, 6460.
35. Bommer, J. C., Morse, K. W., *Chem. Commun.* (1977) 137.
36. Atwood, J. L., Rogers, R. D., Kutal, C., Grutsch, P. A., *Chem. Commun.* (1977) 593.
37. Mann, F. G., Purdie, D., Wells, A. F., *J. Chem. Soc.* (1936) 1503.
38. Jardine, F. H., Rule, L., Vohra, A. G., *J. Chem. Soc. A* (1970) 238.
39. Watts, R. J., Crosby, G. A., *J. Am. Chem. Soc.* (1971) **93**, 3184.
40. Demas, J. N., Adamson, A. W., *J. Am. Chem. Soc.* (1971) **93**, 1800.
41. Ballardini, R., Varani, G., Moggi, L., Balzani, V., Olson, K. R., Scandola, F., Hoffman, M. Z., *J. Am. Chem. Soc.* (1975) **97**, 728.
42. Watts, R. J., Harrington, J. S., Van Houten, J., *J. Am. Chem. Soc.* (1977) **99**, 2179.
43. Creutz, C., Sutin, N., *Inorg. Chem.* (1976) **15**, 496.
44. Wrighton, M. S., Markham, J., *J. Phys. Chem.* (1973) **77**, 3042.
45. Grutsch, P. A., Kutal, C., unpublished data.
46. Lamola, A. A., "Techniques of Organic Chemistry," Vol. 14, Chap. 2, P. A. Leermakers, A. Weissberger, Eds., John Wiley, New York, 1969.
47. Sweet, E. M., Kutal, C., unpublished data.
48. Hautala, R. R., Little, J., Sweet, E. M., *Sol. Energy* (1977) **19**, 503.
49. King, R. B., Sweet, E. M., Hanes, R. M., "Abstracts of Papers, 174th National Meeting, ACS, Chicago, August, 1977, *PETR* 41.

RECEIVED September 20, 1977.

11

Catalysis of Olefin Photoreactions by Transition Metal Salts

ROBERT G. SALOMON

Department of Chemistry, Case Western Reserve University, Cleveland, OH 44106

> *Product evolution and sterochemistry as well as deuterium labeling studies suggest photodissociation of an olefin ligand as a key step in olefin photoreactions catalyzed by Rh(I) chloride. Coordinatively unsaturated rhodium complexes so generated can undergo oxidative addition of an allylic C–H bond of a coordinated olefin, leading to rearrangement, or can abstract hydrogen from solvent, leading to hydrogen transfer reduction. In contrast, the influence of olefin concentration on quantum yield and other observations suggests that both olefin ligands remain coordinated to Cu(I) during photodimerization catalyzed by Cu(I) salts. Various novel carbon skeletal reorganizations that occur upon photolysis of methylene cyclopropanes in the presence of Cu(I) suggest that photocupration, light induced transformation of a η^2-Cu(I)–olefin complex into a η^1-β-Cu(I) carbenium ion intermediate, may be operative in some Cu(I)-promoted photoreactions of olefins.*

When we began our studies, it was known that the salts of two metals, rhodium and copper, have the ability to catalyze photochemical reactions of olefins (1–12). This catalysis is especially interesting since the salts form well-characaterized olefin complexes (13, 14, 15, 16) and since the olefin–metal interaction undoubtedly plays a key role in the photochemical process. Our studies on the mechanisms of these reactions will be discussed. Also, a variety of new types of metal-salt catalyzed photochemical reactions that we recently discovered will be described.

It was known that the dimeric rhodium chloride complex 2 catalyzes the photorearrangement of 1,5-cyclooctadiene (1) to the 1,4 isomer 4 and a bicyclooctene isomer **5** (*4*). Upon close reexamination of this photoreaction, another isomer **6** of bicyclooctene was found as well as a very interesting reduction product, cyclooctene (*7*). During the course of mechanistic investigations on rhodium-catalyzed photorearrangement of olefins, we were led to examine the possibility that similar rearrangements might occur with acyclic dienes. Indeed, upon photolysis of 3,3-dimethyl-1,5-hexadiene (**8**) in the presence of rhodium chloride, olefin isomerization also occurred (*17*).

To explore the evolution of products during the photolysis, the appearance of the products and disappearance of starting material as a function of the time of irradiation was examined. The initial rate of formation of the 1,4-diene **4** is approximately equal but opposite in sign to the rate of disappearance of the starting 1,5-diene **1**, suggesting that the primary photorearrangement product from the 1,5-diene is the 1,4-diene and that the other products are derived from further rearrangement of the 1,4-diene. Indeed, when authentic 1,4-cyclooctadiene is irradiated in the presence of rhodium chloride catalyst, the other rearrangement products are obtained. Moreover, the initial rate of formation of the reduction product **7** is very small. There is a noticeable increase in the rate of its formation after all of the 1,5-diene has been consumed. This rate change clearly indicates that the reduction product arises from the reduction of the 1,4-diene, and probably very little comes from reduction of the 1,5-diene. Parenthetically, the apparent inhibition of reduction by the 1,5-diene is typical for rhodium-promoted transfer hydrogenation (*18*). This inhibition may reflect a requirement for coordination of the hydrogen donor, probably the ether solvent, to rhodium. The mechanism of this reduction will be discussed after considering our studies on the mechanism of the initial allylic rearrangement of the 1,5- to the 1,4-diene. The details of these experiments will not be described since they have been published (*17*). In summary, the rhodium-catalyzed photorearrangement of deuterium labeled 1,5-cyclooctadiene was examined, and suitable crossover and product studies established that the rearrangement of 1,5- to 1,4-diene involves a net allylic 1,3-hydride shift which

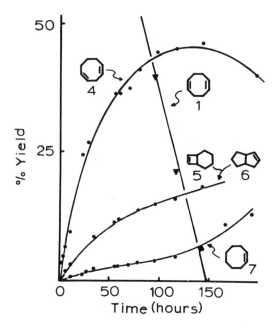

Figure 1. Photolysis of 1,5-cyclooctadiene in the presence of Rh(I)

occurs intramolecularly. Furthermore, the rearrangement shows a deuterium isotope effect indicating that a rate-determining scission of the allylic C–H bond is involved in the process.

There are two quite different mechanisms, vide infra, which might reasonably account for the observed intramolecular hydride shift. These mechanisms are both expected to give the same product, cis,cis-1,4-cyclooctadiene, from the cyclic 1,5-diene. However, an acyclic 1,5-diene could give different products depending on which mechanism were operative.

Therefore, we examined the rhodium-promoted photorearrangement of 3,3-dimethyl-1,5-hexadiene (**8**). One mechanism involves light-induced oxidative addition of an allylic C–H bond to rhodium to give an allylrhodium hydride intermediate **9** which could give isomerized 1,4-diene product **10** by reductive elimination of rhodium. The alternative mechanism involves a light-induced dissociation of one of the two C–C double bonds coordinated to rhodium, leading to a coordinatively unsaturated monoolefin–rhodium complex **11**. The vacant coordination site on rhodium confers exceptional reactivity to this species and would be expected to lead to facile oxidative addition of the allylic C–H bond in a subsequent reaction, not necessarily involving light, which would lead to either a transoid allylrhodium hydride **12** or a cisoid allylrhodium hydride **9**. Reductive elimination of rhodium from these intermediates would lead to either a rearranged *trans-* or a *cis-*1,4-diene product respectively. In fact, when 3,3-dimethyl-1,5-hexadiene is irradiated in the presence of rhodium chloride, both trans- and cis-rearranged products are

obtained. The trans product **13** predominates in 86% yield, accompanied by only 14% of the cis diene **10**. Importantly, the cis and trans dienes are not interconverted under the reaction conditions. Thus, the trans diene is formed directly from the 1,5-diene. This is consistent with the operation of predominantly a photodissociation mechanism for rhodium catalysis of olefin rearrangement. There is at most only a small contribution by direct light-induced formation of **9**.

This mechanism involving generation of a reactive coordinatively unsaturated intermediate also provides an attractive explanation for the production of reduction products during the photoisomerization of olefins in the presence of rhodium. The reduction product from the photolysis of octadeutero cyclooctadiene contains no d^9 or d^{10} material. Thus, hydrogen rather than deuterium is transferred to cyclooctadiene to give reduced product. This hydrogen most likely arises from the diethyl ether that is used as a solvent for these photolyses. The transfer of hydrogen could be achieved by oxidative addition of a C–H bond of ether to the reactive

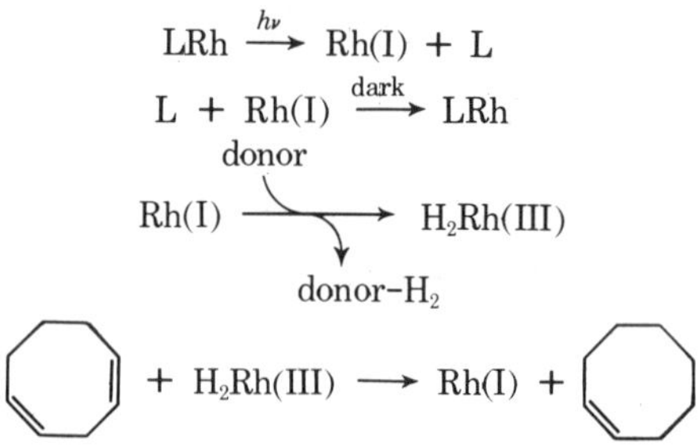

$$LRh \xrightarrow{h\nu} Rh(I) + L$$

$$L + Rh(I) \xrightarrow{dark} LRh$$

$$Rh(I) \xrightarrow{\text{donor}} H_2Rh(III)$$
$$\text{donor-}H_2$$

[cyclooctadiene] $+ H_2Rh(III) \longrightarrow Rh(I) +$ [cyclooctene]

coordinatively unsaturated rhodium complex. Subsequent β-elimination would give a rhodium dihydride that could transfer hydrogen to the diene-regenerating rhodium catalyst. Of particular interest is the fact that the hydrogen transfer in the photochemical reaction occurs at subambient temperatures. Thermal transfer of hydrogen catalyzed by rhodium is indeed known. However, thermal hydrogen transfer is insignificant at temperatures well below 120°C, even for 1,4-dioxane which is a much more effective hydrogen donor than diethyl ether (*18*). There is some reason to be hopeful that the photolytic hydrogen transfer will have some synthetic utility because of the extremely mild conditions under which it can be achieved.

At the outset I noted that both rhodium salts and copper salts were known to catalyze photochemical reactions of organic substrates. The reactions catalyzed by copper are quite different from those catalyzed by rhodium. In contrast to rhodium, copper-catalyzed photochemical reactions of olefins have no analogies in thermal processes. Copper catalyzes a diverse array of novel reactions. Though some mechanistic details have been established, our fundamental understanding of copper catalysis of olefin photoreactions is incomplete.

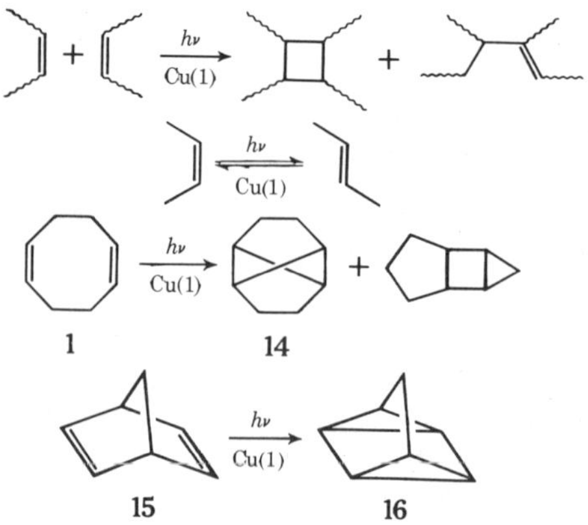

1   14

15   16

In contrast to rhodium, which catalyzes allylic hydrogen migration, copper catalyzes intramolecular photocycloaddition of 1,5-cycloctadiene (1) to give cyclobutane 14 as the major product of photorearrangement (4–9). Photocycloadditions also occur intermolecularly to give cyclobutane dimers (1, 2, 3, 19, 20). Furthermore, norbornadiene (15) efficiently yields cyclobutane 16 upon photolysis in the presence of copper (3, 21, 52, 53). Acyclic olefins as well as cyclooctene and 1,5-cyclooctadiene undergo cis–trans isomerization upon photolysis in the presence of copper (10, 11, 12). Cis–trans isomerization may play a key role in some copper-catalyzed photocycloadditions. Trans–anti-trans cyclobutanes are the major products from photodimerization of cyclohexene and cycloheptene which is catalyzed by cuprous trifluoromethanesulfonate (CuOTf). The cis relationship of the vinyl hydrogens is transformed into a trans relationship for both alkene molecules during photodimerization. This might be attributed to generation of *trans*-cycloalkene which

reacts as a suprafacial partner with a molecule of *cis*-cycloalkene as antarafacial partner in a thermal $2_{\pi a} + 2_{\pi s}$ cycloaddition (20). Cis,trans- and trans,trans-isomers of 1,5-cyclooctadiene (1) have been isolated from photolyses of the cis,cis-isomer in the presence of copper salts and implicated as intermediates in the production of 14 from 1 (11). We recently found that cyclohexene is photosolvated upon irradiation in allyl alcohol solvent in the presence of a copper salt (22). This behavior parallels direct and photosensitized irradiations that involve protonation of transient photogenerated *trans*-cycloalkene (23, 24).

Photodimerization also occurs with olefins that are clearly incapable of generating even transient *trans*-cycloalkene intermediates. For example, endo-dicyclopentadiene (17), which forms the 2:1 complex 18, gives dimer 19 upon irradiation in the presence of Cu(I) (19). Importantly, 17 undergoes intramolecular photocycloaddition upon acetone-sensitized photolysis (25). It is not surprising that intramolecular cycloaddition is facile since two π-bonds in **17** are intimately juxtaposed. It is remarkable that intramolecular cycloaddition is not important in the copper-catalyzed photoreaction. Photodissociation of an olefin ligand is clearly not a viable explanation for the photoactivation of **17** in the presence of Cu(I). Thus, photodissociation of an olefin ligand to generate either an excited olefin or an excited mono-olefin copper complex would be expected to lead to intramolecular photocycloaddition. Apparently, both C–C double bonds must remain coordinated to copper when light is absorbed in order to undergo photodimerization.

Additional evidence which supports the contention that the photodimerization involves reaction of two olefins coordinated to a single copper ion was obtained by studying the effect of olefin concentration on the quantum yield for photodimerization of norbornene (**20**). Thus, the inverse of the quantum yield of both major dimer **21** and minor dimer **22** is correlated linearly with the inverse of the concentration of olefin **20**. A detailed discussion of the interpretation of these results was published

(19). Thus, some mechanism other than generation of a reactive *trans*-cycloalkene, photodissociation, or mere triplet sensitization must be operative in some copper-catalyzed photoreactions.

One possibility we considered is that absorption of light by a copper-olefin complex leads to the addition of the electrophilic copper to the nucleophilic olefin to give a copper-carbenium ion species. The latter

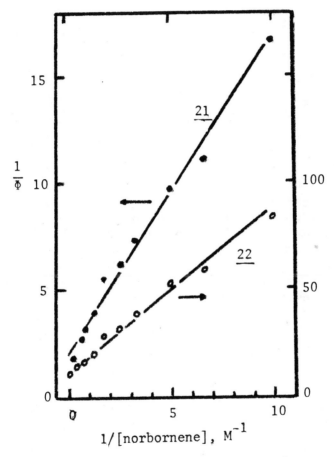

Figure 2. Plot of quantum yields of norbornene dimers at varying concentrations of norbornene with $(CuOTf) = 3.0 \times 10^{-2}M$

can then undergo free rotation about the newly created C–C single bond, followed by elimination of copper leading to trans or cis olefin. Or, the intermediate copper-carbenium ion could react with a second molecule of olefin, leading to a dimeric copper-carbenium ion, which by elimination of copper would produce cyclobutane products. If such a mechanism were operative, then we might expect methylenecyclopropanes to afford cyclopropyl- or cyclopropylcarbinyl-carbenium ions upon irradiation in the presence of copper salts. It is known from solvolysis studies that cyclopropyl- and cyclopropylcarbinyl-carbenium ions undergo a variety of skeletal reorganizations (22–36). Thus, we would expect a cyclopropyl carbenium ion 23 to lead to an allyl carbenium ion intermediate 24 and the cyclopropylcarbinyl ion 25 to afford ring-cleaved homoallyl 26 and ring-expanded cyclobutyl 27 carbenium ions. Products derived from these rearranged carbenium-ion intermediates might be expected.

Indeed, irradiation of methylene cyclopropane 28 or the deuterium-labeled derivative 29 in the presence of Cu(I) resulted in a plethora of skeletal reorganization products including cyclobutenes, allenes, and ring-cleaved vinylcyclohexenes as well as fragmentation products (37). Similarly, for the methylenecyclopropane 30 fused to a seven-membered ring, analogous products as well as, in addition, an isomerized methylene cyclopropane were produced. All of these products are readily understandable in terms of rearrangement of cyclopropylcarbinyl- or cyclopropyl-carbenium ions produced in a key photocupration step leading to the light-induced addition of copper to the C–C double bond. The most interesting of the rearrangements are those giving cyclobutyl products from the cyclopropylcarbinyl precursors. This type of rearrangement has only been observed in reactions involving intermediates in which a vacant orbital is present on the carbinyl carbon of the cyclopropylcarbinyl species, either as cyclopropylcarbinyl-carbenium ions (*30–36*) or cyclopropylcarbinyl carbenes (*38–43*). In particular, cyclopropylcarbinyl radical intermediates do not undergo ring expansion to cyclobutyl derivatives. Thus, a process involving, for example, a triplet intermediate species with radical character on the carbinyl carbon would not be expected to give cyclobutyl products. Neither direct nor sensitized photolyses of methylene cyclopropanes yield cyclobutyl products in the absence of Cu(I) (*44, 45, 46, 47*). Formation of cyclobutyl products **33** and **34** is presumptive evidence for the involvement of cyclopropylcarbinyl-carbenium ions or carbenoids. Furthermore, light-induced cupration may be involved in other olefin photoreactions that are catalyzed by Cu(I). Thus, other products also can be explained by well-precedented transformations of carbenium ions such as cyclopropylcarbinyl to homoallyl rearrangement followed by 1,2-hydride shift and loss of copper to give the allene products **31**. Loss of proton followed by protonolysis of the C–Cu bond might account for the ring-cleaved vinyl-cycloalkenes **32**. Ring opening of the cyclopropyl-carbenium ion to an allyl-carbenium ion and loss of copper might account for the isomeric methylene cyclopropane **35**.

It is noteworthy that a photocupration mechanism also affords an attractive explanation for the formation of some novel minor products from copper-catalyzed olefin photoreactions. For example, the dimeric copper-carbenium ion intermediate **36** could lose copper to produce the cyclobutane products **37** and **38**, but 1,3-hydride shift followed by loss of copper could account for formation of cyclohexyl cyclohexene (39). Similarly, photo-induced cupration of a C–C double bond of 1,5-cyclooctadiene with participation of the neighboring double bond would lead to a bicyclooctyl cation 40 (48). Subsequent loss of copper would give the major cyclobutane product 14, but 1,3-hydride shift, which is particularly favorable in this system, leads to an isomeric copper-carbenium ion 41. Loss of copper from the latter regenerates the catalyst and gives the minor isomeric tricyclooctane product 42. Further studies are needed to test the *trans*-cycloalkene and photocupration hypotheses.

It is tempting to speculate that the preference for photodimerization of two C=C bonds coordinated to the same Cu(I) arises out of the stabilization which is possible for a carbenium-ion intermediate of Structure 43 but not one of Structure 44. That is, a significant stabilizing

interaction is possible between copper and the cationic center in 43 but not in 44. Such an interaction need not involve oxidative addition to Cu(I) with formation of Cu(III). Rather, as was suggested for δ-silver carbenium ions, a less extreme donation of electron density from copper would suffice (49).

Studies of copper catalysis of olefin photoreactions have been limited to alkenes that lack functional groups. Since allyl alcohol is known to form isolable complexes with Cu(I) (50), we examined copper catalysis of olefin photoreactions in allyl alcohol as solvent.

Thus, irradiation of a solution of norbornadiene (15) in allyl alcohol in the presence of Cu(I) results in addition of allyl alcohol to afford the ethers 45 and 46. Since the same products are obtained in high yield from the thermal reaction of quadricyclane (16) with allyl alcohol in the presence of CuOTf, the photochemical reaction probably involves copper-catalyzed photorearrangement of 15 to 16, followed by a dark reaction between 16 and allyl alcohol promoted by CuOTf. Analogous dark reactions of 16 with alcohol promoted by silver salts are well known (51).

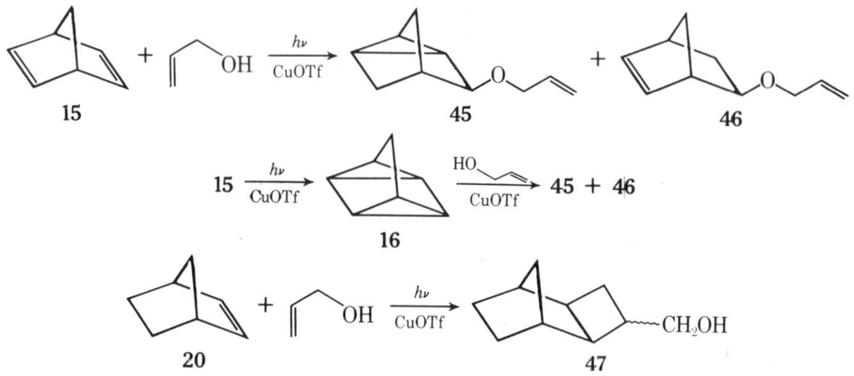

In contrast, irradiation of a 1:1 complex of norbornene (20) and CuOTf in allyl alcohol affords two epimeric cyclobutanes, 47. This observation is especially significant since it demonstrates that photocycloaddition catalyzed by Cu(I) is possible for acyclic olefins. It also shows that photocycloaddition is possible in the presence of a reactive hydroxyl and possibly other functional groups.

Acknowledgment

Acknowledgment is made to the donors of the Petroleum Research Fund, administered by the American Chemical Society, to the Research Corporation, and to the National Science Foundation for support of our research on homogeneous catalysis of organic reactions.

Literature Cited

1. Arnold, D. R., Trecker, D. J., Whipple, E. B., *J. Am. Chem. Soc.* (1965) 87, 2596.
2. Trecker, D. J., Henry, J. P., McKeon, J. E., *J. Am. Chem. Soc.* (1965) 87, 3261.
3. Trecker, D. J., Foote, R. S., Henry, J. P., McKeon, J. E., *J. Am. Chem. Soc.* (1966) 88, 3021.
4. Srinivasan, R., *J. Am. Chem. Soc.* (1964) 86, 3318.
5. Meinwald, J., Kaplan, B. E., *J. Am. Chem. Soc.* (1967) 89, 2611.
6. Srinivasan, R., *J. Am. Chem. Soc.* (1963) 85, 819, 3048.
7. Haller, I., Srinivasan, R., *J. Am. Chem. Soc.* (1966) 88, 5084.

8. Warrener, R. N., Bremner, J. B., *Rev. Pure Appl. Chem.* (1966) **16,** 117.
9. Baldwin, J. E., Greeley, R. H., *J. Am. Chem. Soc.* (1965) **87,** 4514.
10. Nozaki, H., Nisikawa, Y., Kawanisi, M., Noyori, R., *Tetrahedron* (1967) **23,** 2173.
11. Whitesides, G. M., Goe, G. L., Cope, A. C., *J. Am. Chem. Soc.* (1969) **91,** 2608.
12. Deyrup, J. A., Betkouski, M., *J. Org. Chem.* (1972) **37,** 3561.
13. Salomon, R. G., Kochi, J. K., *J. Am. Chem. Soc.* (1973) **95,** 1889.
14. Chatt, J., Venanzi, L. M., *J. Chem. Soc.* (1957) 4735.
15. Ibers, J. A., Snyder, R. G., *J. Am. Chem. Soc.* (1962) **84,** 495.
16. Ibers, J. A., Snyder, R. G., *Acta Crystallogr.* (1962) **15,** 923.
17. Salomon, R. G., El Sandi, N., *J. Am. Chem. Soc.* (1975) **97,** 6214.
18. Nishiguchi, T., Tachi, K., Fukuzumi, K., *J. Am. Chem. Soc.* (1972) **94,** 8916.
19. Salomon, R. G., Kochi, J. K., *J. Am. Chem. Soc.* (1974) **96,** 1137.
20. Salomon, R. G., Folting, K., Streib, W. E., Kochi, J. K., *J. Am. Chem. Soc.* (1974) **96,** 1145.
21. Srinivasan, R., U.S. Patent 3,350,291 (1967) (*Chem. Abstr.* (1968) **68,** 77855u).
22. Salomon, R. G., Cassadonte, D., unpublished data.
23. Marshall, J. A., *Acc. Chem. Res.* (1969) **2,** 33.
24. Kropp, P. J., Reardon, E. J., Jr., Gaibel, Z. L. F., Willard, K. F., Hattaway, J. H., Jr., *J. Am. Chem. Soc.* (1973) **95,** 7058.
25. Schenck, G. O., Steinmetz, R., *Chem. Ber.* (1963) **96,** 520.
26. Skell, P. S., Sandler, S. R., *J. Am. Chem. Soc.* (1958) **80,** 2024.
27. Schweizer, E. E., Parham, W. E., *J. Am. Chem. Soc.* (1960) **82,** 4085.
28. DePuy, C. H., Schanck, L. G., Hausser, J. W., Wiedemann, W., *J. Am. Chem. Soc.* (1965) **87,** 4006.
29. DePuy, C. H., Schanck, L. G., Hausser, J. W., *J. Am. Chem. Soc.* (1966) **88,** 3343.
30. Breslow, R., "Molecular Rearrangements," P. de Mayo, Ed., Vol. I, p. 233, Wiley, New York, 1963.
31. Deno, N. C., *Prog. Phys. Org. Chem.* (1964) **2,** 129.
32. Dewar, M. J. S., Marchand, A. P., *Ann. Rev. Phys. Chem.* (1965) **16,** 321.
33. Servis, K. L., Roberts, J. D., *J. Am. Chem. Soc.* (1965) **87,** 1331.
34. Vogel, M., Roberts, J. D., *J. Am. Chem. Soc.* (1966) **88,** 2262.
35. Schleyer, P. von R., Van Dine, G. S., *J. Am. Chem. Soc.* (1966) **88,** 2321.
36. Majerski, Z., Nikoletić, M., Borčić, S., Sunko, D. E., *Tetrahedron* (1967) **23,** 661.
37. Salomon, R. G., Salomon, M. F., *J. Am. Chem. Soc.* (1976) **98,** 7454.
38. Friedman, L., Shechter, H., *J. Am. Chem. Soc.* (1960) **82,** 1002.
39. Smith, J. A., Shechter, H., Bayless, J., Friedman, L., *J. Am. Chem. Soc.* (1965) **87,** 659.
40. Wilt, J. W., Kosturik, J. M., Orlowski, R. C., *J. Org. Chem.* (1965) **30,** 1052.
41. Wilt, J. W., Zawadzki, J. F., Schultenover, D. G., *J. Org. Chem.* (1966) **31,** 876.
42. Kirmse, W., Pook, K.-H., *Angew. Chem., Int. Ed. Engl.* (1966) **5,** 594.
43. Kirmse, W., Pook, K.-H., *Chem. Ber.* (1965) **98,** 4022.
44. Kende, A. S., Goldschmidt, Z., Smith, R. F., *J. Am. Chem. Soc.* (1970) **92,** 7606.
45. Gilbert, J. C., Butler, J. R., *J. Am. Chem. Soc.* (1970) **92,** 7493.
46. Brinton, R. K., *J. Phys. Chem.* (1968) **72,** 321.
47. Hill, K. L., Doepker, R. D., *J. Phys. Chem.* (1972) **76,** 3153.
48. Anderson, C. B., Burreson, B. J., *Chem. Ind. (London)* (1967) 620.
49. Eaton, P. E., Cassar, L., Hudson, R. A., Hwang, D. R., *J. Org. Chem.* (1976) **41,** 1445.

50. Ishino, Y., Ogura, T., Noda, K., Hirashima, T., Manabe, O., *Bull. Chem. Soc. Jpn.* (1972) **45**, 150.
51. Koser, G. F., Pappas, P. R., Yu, S.-M., *Tetrahedron Lett.* (1973) 4943.
52. Schwendiman, D. P., Kutal, C., *Inorg. Chem.* (1977) **16**, 719.
53. Schwendiman, D. P., Kutal, C., *J. Am. Chem. Soc.* (1977) **99**, 5677.

RECEIVED September 20, 1977.

12

Photocatalyzed Reactions of Alkenes with Silanes Using Trinuclear Metal Carbonyl Catalyst Precursors

RICHARD G. AUSTIN, RALPH S. PAONESSA, PAUL J. GIORDANO, and MARK S. WRIGHTON[1]

Department of Chemistry, Massachusetts Institute of Technology, Cambridge, MA 02139

> $M_3(CO)_{12}$ ($M =$ Fe, Ru, Os) are effective photocatalysts at $298°K$ for alkene isomerization and for reaction of alkenes with silanes; detailed studies on 1-pentene isomerization and 1-pentene reaction with $HSiEt_3$ reveal a number of trends. Isomerization activity depends on M in the order $Fe > Ru > Os$. Photocatalyzed reaction of 1-pentene and $HSiEt_3$ yields n-pentane, (n-pentyl)$SiEt_3$, and three isomers of (pentenyl)$SiEt_3$ where all silicon-containing products result from addition of the -$SiEt_3$ moiety to the terminal carbon of the hydrocarbon. The distribution of silicon-containing products depends on M and the ratio of $HSiEt_3$ and 1-pentene; $Os_3(CO)_{12}$ gives predominantly the (n-pentyl)$SiEt_3$ while $Fe_3(CO)_{12}$ and $Ru_3(CO)_{12}$ give mainly isomers of (pentenyl)$SiEt_3$. Higher 1-pentene:$HSiEt_3$ ratios give higher yields of (pentenyl)$SiEt_3$. Preliminary findings are consistent with photogeneration of mononuclear catalysts from $M =$ Fe or Ru, while for $Os_3(CO)_{12}$ the cluster may remain intact.

Electronic excited states of certain low valent organometallic complexes are known to relax to ground state species that are coordinatively unsaturated (1, 2, 3, 4, 5). Photoinduced ligand dissociation (Reaction 1) from numerous mononuclear complexes is believed to reflect the intermediacy of reactive ligand field excited states; i.e., the states that

---

[1] Address correspondence to this author.

$$L_xM(CO)_n \xrightarrow{h\nu} L_xM(CO)_{n-1} + CO \qquad (1a)$$

$$\xrightarrow{h\nu} L_{x-1}M(CO)_n + L \qquad (1b)$$

$$\text{\Large$>$}M'-M\text{\Large$<$} \xrightarrow{h\nu} \text{\Large$>$}M'\cdot + \cdot M\text{\Large$<$} \qquad (2)$$

are strongly sigma-antibonding with respect to the metal–ligand interaction (6, 7, 8). A large class of metal–metal bonded dinuclear complexes fragment (Reaction 2) subsequent to the population of the lowest lying excited states which are destabilizing with respect to the metal–metal sigma bonding (9–15). Reaction 1 leads to the generation of $16e^-$ species from $18e^-$ precursors, and Reaction 2 results in the formation of $17e^-$ metal radicals from diamagnetic precursors. Photoinduced reductive elimination of hydrogen from di- and polyhydride complexes (Reaction 3) provides another photochemical entry into coordinatively unsaturated species (16, 17, 18).

$$L_nM\begin{matrix}H\\ \\H\end{matrix} \xrightarrow{h\nu} L_nM + H_2 \qquad (3)$$

Since coordinatively unsaturated species are believed to play a key role in homogeneous catalysis (19, 20, 21), there exists the possibility of initiating, accelerating, and altering homogeneous catalysis by optical irradiation (22). Photochemical routes to active catalysts from thermally inert precursors may allow more convenient handling of the organometallic species, and control of the rate of catalytic processes may be achieved simply by variation in light intensity. However, in fundamental terms, the real interest may rest in the fact that photochemically synthesized catalysts can result from direct decay of an electronic excited state. Such being the case, one can hold out the promise that the photogenerated catalyst may be one that is unique and only preparable by photochemical means.

A large number of organometallic substances are known catalysts for organic reactions, and photocatalytic schemes for certain reactions have been reported in the literature. So far, the main examples of photocatalysis appear to involve Reaction 1 as the key photochemical step in the process, and the single most important advantage that has been found generally is that reaction can be sustained at lower temperatures than in the thermal process. For example, in the $Fe(CO)_5$-catalyzed hydrogenation of olefins, loss of CO from $Fe(CO)_5$ is the rate limiting step and requires high temperatures (23, 24, 25, 26, 27). However, irradiation of Fe(CO)$_5$ is known to result in the efficient dissociative loss of CO (1, 28, 29) (Reaction 4) and it has been shown that irradiation of Fe(CO)$_5$ in the presence of hydrogen and an olefin can lead to hydro-

$$Fe(CO)_5 \xrightarrow{h\nu} Fe(CO)_4 + CO \qquad (4)$$

genation under conditions where no thermal reaction obtains (30). Photocatalysis at low temperatures is important for at least three reasons. First, catalytic chemistry of thermally sensitive molecules may be possible. Second, compared with the thermal process, there may be a different rate-limiting step in the thermal events subsequent to photochemical catalyst generation, allowing more selective catalytic chemistry. Third, the generation of a catalytically active species under low temperature conditions may allow the characterization of intermediates that generally are not detectable at higher temperatures.

At the present time there are examples (22) of photocatalyzed olefin isomerization (30, 31, 32), hydrogenation (30, 33–37), hydrosilation (40, 41), oligomerization (42, 43), and metathesis (44, 45, 46). These involve, presumably, photochemistry like that in Reaction 1, followed by low activation-energy thermal steps that parallel known catalytic chemistry. The observation that irradiation of $(\eta^5\text{-}C_5H_5)_2WH_2$ (47) or $(\eta^5C_5H_5)_2$-W(CO) (48) leads to oxidative addition of C_6H_6 (Reaction 5) does illustrate that very reactive intermediates can be generated by photoexcitation, and we can expect that this area is to be one of active pursuit in the future.

(5)

Catalysis involving the fragments from the photoinduced cleavage of metal–metal bonds has not received much attention yet, but numerous photosensitive di- and polynuclear clusters are now known (9–15), and studies of catalytic chemistry seem appropriate. For a large number of dinuclear compounds, homolytic cleavage of the metal–metal bond is a very efficient process occurring from the lowest excited states, and dissociative loss of ligands is at best a minor component of the decay paths (9–15). For the trinuclear clusters, $M_3(CO)_{12}$, (M = Fe, Ru, Os), there have been reports of the isolation of mononuclear, dinuclear, and trinuclear products from irradiation of $M_3(CO)_{12}$ in the presence of nucleophiles or oxidative addition substrates (49–58). Consequently, we have undertaken studies directed towards assessing the catalytic activity of the intermediates in the photochemical reactions of $M_3(CO)_{12}$.

We have chosen to investigate the catalytic reactions resulting from irradiation of $M_3(CO)_{12}$ in the presence of alkenes or alkenes and silicon hydrides. This seems to be a reasonable starting point since $Fe(CO)_5$-photocatalyzed reactions of alkenes and alkenes and silanes have been reported (41). Further, both $Ru_3(CO)_{12}$ and $Os_3(CO)_{12}$ result in the formation of mono- and dinuclear oxidative addition products when irradiated in the presence of a silicon hydride (51, 52, 53, 54). Irradiation of $Os_3(CO)_{12}$ in the presence of 1,5-cyclooctadiene has been reported to yield some $(1,3\text{-cyclooctadiene})Os(CO)_3$, evidencing an ability to isomerize an olefin (59). There is an early report of the $Fe_3(CO)_{12}$- and $Os_2(CO)_9$-photocatalyzed isomerization of 1-undecene (60). Both Fe_3-$(CO)_{12}$ (60–65) and $Ru_3(CO)_{12}$ (66) are known catalysts for alkene isomerization, and recently there have been a number of interesting reports concerning chemistry of $M_3(CO)_{12}$ and derivatives with olefins (see Ref. 67–79).

Results

Absorption Spectra of $M_3(CO)_{12}$. The optical absorption spectra of the $M_3(CO)_{12}$ complexes are given in Figure 1, and the band positions and intensities are summarized in Table I. There are a number of fairly intense absorptions for each complex, but the noteworthy trend is that the first absorption system position is in the order Os > Ru > Fe. Thus, low energy visible excitation is possible only for the iron complex.

Isomerization of Alkenes. Each of the $M_3(CO)_{12}$ complexes is effective with respect to photocatalyzed alkene isomerization. Reaction 6 has

$$\diagup\!\!\!\diagdown\!\!\!\diagup\!\!\!\diagdown \xrightarrow[\substack{M_3(CO)_{12}(\sim 10^{-3}M) \\ 298\ °K}]{h\nu} \diagup\!\!=\!\!\diagdown\!\!\!\diagdown + \diagup\!\!\!\diagdown\!\!\!=\!\!\!\diagdown\!\!\!\diagdown \quad (6)$$

been investigated and Table II summarizes the key findings. The general trend is that the effectiveness of the isomerization (observed rate and extent conversion) seems to follow the ordering Fe > Ru >> Os. Indeed, only $Fe_3(CO)_{12}$ seems to bring the linear pentenes to their thermodynamic ratio (72) in short times. None of the $M_3(CO)_{12}$ complexes gives

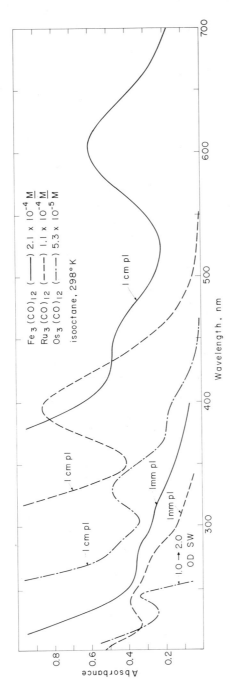

Figure 1. Optical absorption spectra of $M_3(CO)_{12}$ complexes at 298°K in isooctane solution; cf. Table I for band maxima and molar absorptivities

Table I. Spectral Properties of $M_3(CO)_{12}$[a]

M	Bands, nm (ϵ, lmol^{-1} cm^{-1})
Fe	603 (2900)
	440 sh (2380)
	315 sh (12,400)
	275 sh (17,700)
	192 (> 70,000)
Ru	395 (7700)
	268 sh (27,000)
	239 (35,500)
	203 sh (48,000)
Os	385 sh (3700)
	329 (9300)
	288 sh (8500)
	244 (26,00)

[a] Isooctane solution at 298°K, cf. Figure 1.

Table II. $M_3(CO)_{12}$-Photocatalyzed Alkene Isomerization[a]

M	Irrdn. Time (hr)	%1-pentene	%trans-2-pentene	%cis-2-pentene
Fe	0	100	—	—
	1	7.0	74.4	18.6
	12	3.0	76.0	21.0
Ru	0	100	—	—
	1	70.0	24.7	5.1
	24	61.0	34.0	5.0
Os	0	100	—	—
	1	> 99	< 1	< 1
	17	95.8	3.3	1.0

[a] Irradiation of 1 mL of $10^{-3}M$ $M_3(CO)_{12}$, $2M$ 1-pentene degassed benzene solutions. Irradiation at 298°K with GE Black Light through borosilicate glass.

thermal reaction on the time scale of the photochemical experiments that are carried out at 298°K.

Quite importantly, $Fe_3(CO)_{12}$ photocatalyzes the pentene isomerization upon excitation with low energy visible light (Table III). The reaction is accompanied by the disappearance of $Fe_3(CO)_{12}$ and the formation of mononuclear iron carbonyl species including $Fe(CO)_5$ and $Fe(CO)_4$(pentene). Such species have been identified as products by their characteristic CO stretching frequencies in the IR. Data in Table III show that the number of alkene molecules isomerized per iron atom initially present is quite large. It is apparent that the linear pentenes can be equilibrated to the thermodynamic mixture (*80*) by the visible light photocatalysis procedure.

Reaction of Alkenes with Silanes. Irradiation of $M_3(CO)_{12}$ in the presence of 1-pentene and $HSiEt_3$ proceeds generally according to Reaction 7. The three (pentenyl)$SiEt_3$ products **I**, **II**, and **III**, are the major

$$\diagup\!\!\diagdown\!\!\diagup + HSiEt_3 \xrightarrow[M_3(CO)_{12}(\sim 10^{-3}M)]{h\nu} n\text{-}C_5H_{12} + (n\text{-pentyl})SiEt_3 +$$

$$Et_3Si\diagdown\!\!\diagup\!\!\diagdown + Et_3Si\diagdown\!\!\diagup\!\!\diagdown + Et_3Si\diagdown\!\!\diagup\!\!\diagdown \quad (7)$$

$$\qquad\qquad \mathbf{I} \qquad\qquad\qquad \mathbf{II} \qquad\qquad\qquad \mathbf{III}$$

products but sometimes are accompanied by trace amounts of what appear to be other isomers (cis–trans and hydrogen-shift products). The products have been identified by their mass spectra and by comparison with authentic samples (VPC retention time and mass spectrum). The photocatalysis can be carried out on neat mixtures of the 1-pentene and $HSiEt_3$, and consumption of the limiting reagent generally exceeds 90%.

Table III. $Fe_3(CO)_{12}$-Photocatalyzed Pentene Isomerization with 550-nm Excitation[a]

Starting Alkene [M]	Irrdn. Time (min)	%1-pentene	%trans-2-pentene	%cis-2-pentene
1-pentene [1.7]	0	99.7	0.2	0.1
	108	79.2	16.3	4.5
	230	59.3	31.8	8.9
	465	29.0	54.4	16.6
	810	3.6	76.0	20.0
trans-2-pentene [0.7]	0	—	> 99.0	—
	240	1.0	97.1	1.8
	820	2.3	91.4	6.3
	2670	3.4	73.2	23.4
cis-2-pentene [0.7]	0	—	—	> 99.0
	180	2.0	12.9	85.1
	790	2.8	40.3	56.9
	2640	2.9	59.4	37.8

[a] Degassed benzene solutions (3.0 mL) of $10^{-3}M$ $Fe_3(CO)_{12}$ and alkene were irradiated with 550-nm output from 550-W medium pressure Hg lamps (Hanovia) at 298°K in a merry-go-round; 10^{-7}-10^{-6} ein/min incident on the sample.

Table IV. $M_3(CO)_{12}$-Photocatalyzed

Catalyst Precursor	Irrdn. Time	% Conversion	%n-C_5H_{12}
$Fe_3(CO)_{12}$	5 min	2	49.0
	1 hr	15	47.4
	18 hr	80	44.4
$Ru_3(CO)_{12}$	1 hr	15	48.4
	2 hr	30	48.7
	24 hr	96	46.1
$Os_3(CO)_{12}$	1 hr	24	13.4
	24 hr	> 99	15.0

a Neat, 1:1 mole ratio of 1-pentene and $HSiEt_3$. One mL degassed solutions of

Conversion and product distribution as a function of irradiation time are detailed in Table IV, and the data show that the distribution of products is such that the amount of n-C_5H_{12} is about the same as the amount of the (pentenyl)$SiEt_3$ isomers combined. Further, each $M_3(CO)_{12}$ gives its own characteristic ratio of products, and the distribution of the (pentenyl)$SiEt_3$ isomers is fairly constant through the course of the reaction.

Comparison of $Fe(CO)_5$ and $Fe_3(CO)_{12}$. Table V shows a comparison of the silicon-containing product distribution using $Fe(CO)_5$ or $Fe_3(CO)_{12}$ as the catalyst precursor. First, note that either 550-nm or near-UV excitation of the $Fe_3(CO)_{12}$ gives the same initial distribution of

Table V. Comparison of $Fe(CO)_5$- and $Fe_3(CO)_{12}$-Photocatalyzed Reaction of 1-Pentene with $HSiEt_3$ *a*

Catalyst Precursor	Irrdn. λ (nm) *b*	% Conv.	Product Distribution			
			(n-pentyl)-$SiEt_3$	(pentenyl)$SiEt_3$		
				I	II	III
$Fe(CO)_5$	355	2	16.5	21.3	52.3	9.9
		> 80	17.5	16.1	51.2	15.2
$Fe_3(CO)_{12}$	355	2	6.1	20.2	62.9	10.9
		30	9.1	20.3	58.9	11.7
		80	15.9	17.2	51.7	15.1
$Fe_3(CO)_{12}$	550	1	4.8	17.5	66.2	11.5
		4	6.5	18.6	64.3	10.6
		26	8.2	20.7	60.4	10.6

a One mL samples of $10^{-3}M$ catalyst precursor in degassed 1:1 mole ratio of 1-pentene and $HSiEt_3$.
b 355-nm irradiation was with a GE Black Light, and the 550 nm irradiation was with a filtered 550-W Hanovia medium pressure Hg lamp.

Reaction of 1-Pentene and HSiEt₃ [a]

%(n-pentyl)SiEt₃	%(pentenyl)SiEt₃		
	I	II	III
4.2	8.8	33.0	4.9
4.5	9.2	33.6	5.3
8.9	9.6	28.8	8.4
3.5	42.8	4.6	~0.6
3.2	40.8	5.7	1.5
2.8	45.2	4.3	1.5
69.8	13.9	2.9	<1
63.2	17.7	4.1	<1

$10^{-3}M$ $M_3(CO)_{12}$ irradiated with GE Black Light at 298°K.

products. However, we do note some tendency for the 550-nm excitation to give a smaller extent conversion than for near-UV excitation. Second, the distribution of products for $Fe(CO)_5$ and $Fe_3(CO)_{12}$ is very similar. In particular, the distribution of products I, II, and III is nearly the same, although there does seem to be an experimentally significant, but small, variation in the (n-pentyl)SiEt₃ to (pentenyl)SiEt₃ ratio. One final comparison between $Fe(CO)_5$ and $Fe_3(CO)_{12}$ is appropriate here. Table III shows the initial ratio of cis- and trans-2-pentene formed from $Fe_3(CO)_{12}$-photocatalyzed 1-pentene isomerization. This ratio is essentially the same as that reported previously for $Fe(CO)_5$ (30).

Product Distribution Variation with Alkene:Silane Ratio. Variation in the initial ratio of 1-pentene to HSiEt₃ gives large variations in the distribution of products. Table VI gives some data showing that with an excess of the 1-pentene, there is a greater tendency to form the (pentenyl)SiEt₃ products. The effect is particularly striking for $Fe_3(CO)_{12}$ where the (n-pentyl)SiEt₃ is a very minor component of the product mixture at an initial 10:1 1-pentene:HSiEt₃ ratio. At the other extreme, $Os_3(CO)_{12}$, which gives a substantially larger fraction of the (n-pentyl)SiEt₃, gives almost exclusively that product at the 1:10 ratio of 1-pentene:HSiEt₃. Curiously, $Ru_3(CO)_{12}$ is relatively insensitive to substrate ratio, giving mainly (pentenyl)SiEt₃ products under all conditions. It is worth noting that there do seem to be some relatively minor changes in the distribution of isomeric (pentenyl)SiEt₃ products as the substrate ratio is changed.

Selectivity for Terminal Alkene. All three $M_3(CO)_{12}$ complexes apparently only result in products from reaction of 1-pentene. The evidence for this is several-fold. First, all of the silicon-containing products have the -SiEt₃ moiety bonded to the terminal carbon of a linear C_5 fragment. By exposure to photocatalysis conditions in separate experi-

Table VI. Variation in $M_3(CO)_{12}$ Photocatalysis Product Distribution with Variation in Alkene:Silane Ratio[a]

M	1-pentene: HSiEt$_3$[b]	% Conversion[c]	% Alkylsilane	% Alkenylsilanes		
				I	II	III
Fe	10:1	5	<1	18.2	70.3	11.4
		>80	<1	17.6	60.0	22.3
	1:1	2	8.2	17.3	64.9	9.6
		>80	15.9	17.2	51.8	15.1
	1:10	5	37.2	13.0	40.9	8.7
		>80	52.3	12.5	25.5	9.6
Ru	10:1	15	5.3	84.4	10.3	<1
		>90	2.7	66.8	24.5	6.0
	1:1	15	6.9	84.7	9.3	<1
		>90	5.3	83.8	8.0	3.0
	1:10	5	5.2	84.1	10.6	<1
		>80	7.4	82.0	8.3	2.3
Os	10:1	15	62.5	32.0	5.5	<1
		>80	58.6	34.1	7.3	<1
	1:1	24	80.6	15.7	3.3	<1
		>90	74.4	20.8	4.8	<1
	1:10	15	93.4	5.8	<1	<1
		>80	83.3	13.1	2.8	<1

[a] Irradiation of 1 mL $10^{-3}M$ $M_3(CO)_{12}$ degassed solutions in borosilicate glass ampules at 298°K with GE Black Light.
[b] Mole ratio of alkene and silane.
[c] Based on limiting reagent.

ments, there is no evidence that the possible internal products are unstable to the photocatalysis conditions. Second, the 1-pentene in a mixture of linear pentenes can be consumed completely by the photocatalyzed reaction with HSiEt$_3$ before there is any substantial reaction of the cis-

Table VII. $M_3(CO)_{12}$-Photocatalyzed Reaction of HSiEt$_3$ and cis-2-Pentene[a]

M	Irrdn. Time (hr)	% Conversion	%(n-pentyl)-SiEt$_3$	%(pentenyl)SiEt$_3$		
				I	II	III
Fe	18	17	23.9	7.7	53.6	14.8
Ru	18	5	7.4	84.9	7.7	<1
Os	18	~2	75.7	24.3	<1	<1

[a] Irradiation of 1.0 mL degassed solutions of $10^{-3}M$ $M_3(CO)_{12}$ in 1:1 cis-2-pentene: HSiEt$_3$ solutions. Irradiation was carried out at 298°K with a GE Black Light.

or *trans*-2-pentene. This fact is illustrated clearly in Figure 2 which shows the gas chromatographic traces of a pentene mixture as a function of $Fe_3(CO)_{12}$ photocatalysis time. Ultimately, more of the pentene can be consumed by the reaction with $HSiEt_3$, albeit at a much slower rate. Finally, attempted reaction of pure *cis*-2-pentene with $HSiEt_3$ by the $M_3(CO)_{12}$ photocatalysis procedure results in very slow rates of consumption in comparison with the 1-pentene reaction. The extent conversion

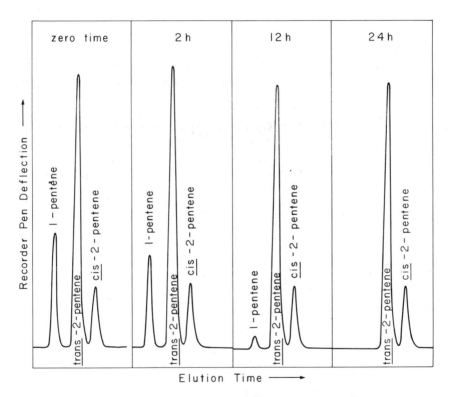

Figure 2. Gas chromatographic traces showing pentene distribution as a function of photocatalysis reaction time using $Fe_3(CO)_{12}(10^{-3}M)$. Near-UV excitation was used at 298°K. The initial solution was a neat solution 1:1, alkene: $HSiEt_3$. Equal sized injections were made at each time.

after 18 hr of irradiation is given in Table VII. Note that the extent of conversion correlates with the effectiveness of the isomerization activity of $M_3(CO)_{12}$; but in each case, the extent conversion is much less than when 1-pentene is the starting alkene.

Relative Rates of Isomerization and Hydrosilation. Some of the data which shows that the terminal alkene reacts selectively with $HSiEt_3$ reveal that the photocatalysis procedure does not result in the rapid equilibration of the linear pentenes under the reaction conditions. An analysis of the unreacted pentene at various stages in the reaction with HSiEt₃ shows that alkene isomerization is slow or, at best, competitive with the reaction to give silicon-containing products (Table VIII). The effect is particularly striking for $Os_3(CO)_{12}$, which shows little or no isomerization activity at any stage in the reaction. But even for $Fe_3(CO)_{12}$, where pentene isomerization is very effectively photocatalyzed in the absence of HSiEt₃, we observe relatively slow isomerization. This statement is conclusive because we have shown that 1-pentene can be removed from a thermodynamic mixture of the linear pentenes by reaction with HSiEt₃ (Figure 2).

The relatively slow pentene isomerization is consistent, too, with the observation that the distribution of isomeric (pentenyl)SiEt₃ products is

Table VIII. $M_3(CO)_{12}$-Photocatalyzed Isomerization vs. Hydrosilation[a]

M	% Consumption of Alkene[b]	% trans-2-pentene	% cis-2-pentene
Fe	0	0.0	0.0
	8.0	5.1	1.0
	> 90	77.4	22.6
Ru	> 80	28.0	3.2
Os	20	< 1	< 1
	> 90	2.0	< 1

[a] Solution is initially 1:1 mole ratio of HSiEt₃ and 1-pentene with $10^{-3}M$ $M_3(CO)_{12}$. One mL degassed samples in borosilicate glass ampules at 298°K were irradiated with a GE Black Light.
[b] Products are alkyl- and alkenylsilanes.

essentially independent of percent conversion. Since each $M_3(CO)_{12}$ complex gives a different ratio of **I, II,** and **III**, it is evident that equilibration of these olefinic products generally is not efficient under the reaction conditions.

Photocatalysis Quantum Yields. Data in Table IX show representative quantum yields for $M_3(CO)_{12}$-photocatalyzed 1-pentene isomerization and 1-pentene reaction with HSiEt₃. The quantum yields are defined here to be the number of alkene molecules reacted per photon incident on the sample. Near-UV (355 nm ± 25 nm) irradiation was used. The noteworthy finding here is that in all cases but one, the quantum yield is significantly greater than unity. This fact allows the definitive conclusion that irradiation of $M_3(CO)_{12}$ produces catalytically active intermediates whose activity persists for a number of catalytic cycles.

Table IX. Observed Reaction Quantum Yields for
$M_3(CO)_{12}$-Photocatalyzed Reactions

M	Solution	Φ Disappearance[a] of 1-Pentene
Fe	Neat 1-pentene	61
	1-pentene:$HSiEt_3$ (1:1)	16
Ru	Neat 1-pentene	34
	1-pentene:$HSiEt_3$ (1:1)	24
Os	Neat 1-pentene	<1
	1-pentene:$HSiEt_3$ (1:1)	13

[a] Disappearance of 1-pentene measured as a function of irradiation time (2.2 × 10^{-6} ein/min at 355-nm incident on sample). In neat 1-pentene, the products are cis- and trans-2-pentene and in the 1-pentene:$HSiEt_3$ solutions, the disappearance of all alkenes was monitored.

$M_3(CO)_{12}$ **Photochemistry.** The $M_3(CO)_{12}$ complexes are quite evidently photosensitive, and excitation apparently leads to intermediates capable of alkene/silane chemistry. We have begun to characterize the primary isolable photoproducts from irradiation of $M_3(CO)_{12}$ to gain insight into the nature of the reactive species. As irradiation of $M_3(CO)_{12}$ in the presence of silicon hydrides or nucleophiles already has been shown to give mono-, di-, and trinuclear products (*49–59*), we speculate that there are two possible primary photoreactions and several secondary thermal pathways. Reactions 8 and 9 seem to be the two possible results

$$M_3(CO)_{12} \underset{}{\overset{h\nu}{\rightleftarrows}} M_3(CO)_{11} + CO \qquad (8)$$

$$M_3(CO)_{12} \overset{h\nu}{\rightleftarrows} (OC)_4M \overset{M(CO)_4}{\diagdown} M(CO)_4 \qquad (9)$$

of decay of the excited state(s). Reaction 8 seemingly would result in simple CO substitution with a nucleophile L (Reaction 10), but the

$$M_3(CO)_{11} \xrightarrow{L} M_3(CO)_{11}L \qquad (10)$$

diradical product in Reaction 9 could give the same product since it has been shown that 17e centers are coordinatively labile (*9, 10, 81, 82, 83*). A possible route to simple substitution through the diradical is as shown in the sequence of Reactions 9, 11, 12, and 10.

$$(OC)_4M \overset{M(CO)_4}{\diagdown} M(CO)_4 \xrightarrow{-CO} (OC)_4M \overset{M(CO)_4}{\diagdown} M(CO)_3 \qquad (11)$$

$$(OC)_4M \overset{M(CO)_4}{\diagdown} M(CO)_3 \longrightarrow M_3(CO)_{11} \qquad (12)$$

Table X. IR Spectral

Complex	Solvent
Fe(CO)$_5$	isooctane
Fe$_3$(CO)$_{12}$	n-hexane
Fe$_3$(CO)$_{11}$PPh$_3$	cyclohexane
Fe(CO)$_4$PPh$_3$	isooctane
Fe(CO)$_3$(PPh$_3$)$_2$	isooctane
Fe(CO)$_4$(C$_2$H$_4$)	—
Fe(CO)$_4$(1-pentene)	isooctane
cis-HFe(CO)$_4$SiEt$_3$	isooctane
cis-HFe(CO)$_4$SiMe$_3$	isooctane
Ru(CO)$_5$	heptane
Ru$_3$(CO)$_{12}$	isooctane
Ru$_3$(CO)$_9$(PPh$_3$)$_3$	cyclohexane
Ru(CO)$_4$PPh$_3$	heptane
Ru(CO)$_3$(PPh$_3$)$_2$	methylcyclohexane
Ru(CO)$_4$(pentene)	isooctane
Ru(CO)$_4$(ethylene)	heptane
Os(CO)$_5$	heptane
Os$_3$(CO)$_{12}$	isooctane
Os$_3$(CO)$_{11}$PPh$_3$	carbon tetrachloride
Os$_3$(CO)$_{10}$(PPh$_3$)$_2$	carbon tetrachloride
Os$_3$(CO)$_9$(PPh$_3$)$_3$	carbon tetrachloride
Os(CO)$_4$PPh$_3$	heptane
Os(CO)$_3$(PPh$_3$)$_2$	tetrahydrofuran
Os$_3$(CO)$_{10}$H$_2$	cyclohexane

The photogenerated intermediate(s) must be capable of chemistry other than simple substitution, however. This conclusion is reached by noting that irradiation of Fe$_3$(CO)$_{12}$ or Ru$_3$(CO)$_{12}$, but interestingly not Os$_3$(CO)$_{12}$, under CO gives the corresponding M(CO)$_5$ species. This was reported previously for M = Ru (49), and we have extended this to M = Fe. Likewise, visible (not absorbed by M(CO)$_5$) irradiation of M$_3$(CO)$_{12}$ (M = Fe, Ru) gives M(CO)$_4$(pentene) in the presence of pentene, and some cis-HFe(CO)$_4$SiEt$_3$ results upon 633-nm irradiation of the iron cluster in the presence of HSiEt$_3$. For M = Os, under the same conditions as for Fe or Ru (298°K, ~ 1 atm CO), we were unable to detect the formation of Os(CO)$_5$. IR bands in the CO stretching region were used to identify photoproducts, and these bands are included in Table X.

Irradiation of the M$_3$(CO)$_{12}$ species in the presence of PPh$_3$ provides some interesting results. For M = Fe or Ru, the photochemistry at 298°K appears to proceed as indicated in Reaction 13. Plots of the formation of either of the mononuclear products against time (Figure 3)

Data for Relevant Complexes

IR Bands, cm^{-1}	Ref.
2025(s); 2000(vs)	30
2046(s); 2023(m); 2013(sh); 1867(vw); 1835(w)	97
2088(m); 2034(s); 2013(vs); ~1965(w, sh); 1825(vw, br)	84
2054(s); 1978(m); 1942(s)	30
1893(s)	30
2088; 2007; 2013; 1986	98
2084; ; ; 1978	30
2093(w); 2027(m); 2019(s); 2006(s)	This work
2090(m); 2025(m); 2015(s); 2000(s)	41
2035(s); 1999(vs)	99
2061(vs); 2031(s); 2012(m)	This work
2046(vw); 1978(sh); 1970(s); 1933(s); 1929(s); 1920(sh)	100
2060(vs); 1986(m); 1953(vs); 1900(s)	92
1900(s)	85
2103(w); 2018(s); 1990(m)	This work
2104(m); 2021(vs); 1995(s)	49
2034(s); 1991(vs)	99
2070(vs); 2037(vs); 2017(m); 2017(m)	This work
2018(m); 2055(s); 2035(ms); 2019(s); 2000(m); 989(m); 1989(m); 1978(m); 1956(mw)	90
2085(mw); 2030(s); 2012(m); 1998(s); 1969(m); 1951(mw)	90
2053(w); 1999(sh); 1990(m); 1976(s); 1944(m)	90
2060(s); 1980(m); 1943(vs)	92
1890	92
2110(vw); 2076(vs); 2062(s); 2025(vs); 2009(vs); 1987(m); 1969(vw); 1956(vw)	93

$$M_3(CO)_{12} \xrightarrow[PPh_3]{h\nu} M(CO)_4PPh_3 + M(CO)_3(PPh_3)_2 \quad (13)$$

298°K; M=Fe, Ru; alkane solution

show them to be primary photoproducts. For M= Fe, the possible Fe$_3$(CO)$_{11}$(PPh$_3$) is thermally unstable (84) and does give reaction with PPh$_3$ to produce Fe(CO)$_4$PPh$_3$ and Fe(CO)$_3$(PPh$_3$)$_2$. However, for M = Ru, the substituted clusters remain intact in the presence of PPh$_3$ at 298°K (85, 86, 87, 88, 89). We do not see any substantial evidence for the generation of the substituted ruthenium clusters upon near-UV irradiation of Ru$_3$(CO)$_{12}$ in the presence of PPh$_3$. In considerable contrast to the iron and ruthenium results, Os$_3$(CO)$_{12}$ does not appear to give mononuclear complexes as primary products upon near-UV irradiation in the presence of PPh$_3$. Rather, the photochemistry seems to occur as in Reaction 14. The substituted osmium clusters indicated as products are known (90, 91) compounds, as are the mononuclear species M(CO)$_4$-

$$\text{Os}_3(\text{CO})_{12} \xrightarrow[\text{PPh}_3]{h\nu} \text{Os}_3(\text{CO})_n(\text{PPh}_3)_{12-n} \quad (14)$$
$$n = 11, 10, 9$$

PPh$_3$ and M(CO)$_3$(PPh$_3$)$_2$ (92). Again the IR spectral features are conclusive, and the bands are given in Table X.

Finally, with respect to the photochemistry of Os$_3$(CO)$_{12}$, irradiation under hydrogen gives chemistry according to Reaction 15. Not incidentally, irradiation of Os$_3$(CO)$_{12}$ under hydrogen in the presence of 1-pen-

$$\text{Os}_3(\text{CO})_{12} \xrightarrow[\text{isooctane}]{h\nu, \text{H}_2 \text{ (10 psi)}} \text{Os}_3(\text{CO})_{10}\text{H}_2 + 2\text{CO} \quad (15)$$

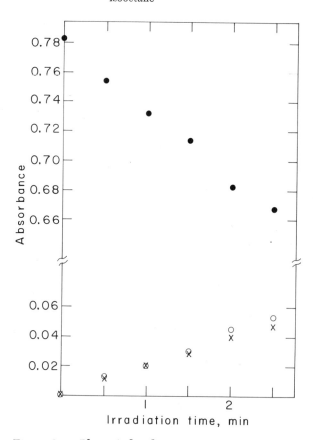

Figure 3a. Plots of absorbance against irradiation time at 2046 cm^{-1} (●) associated with Fe$_3$(CO)$_{12}$, 1942 cm^{-1} (○) associated with Fe(CO)$_4$PPh$_3$, and 1893 cm^{-1} (×) associated with Fe(CO)$_3$(PPh$_3$)$_2$. Spectral changes are for 633-nm irradiation of 3.7 × 10^{-4}M Fe$_3$(CO)$_{12}$, 0.096M PPh$_3$ in isooctane under nitrogen.

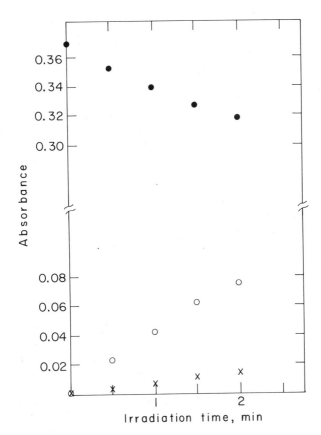

Figure 3b. Plots of absorbance against irradiation time at 2031 cm^{-1} (●) associated with $Ru_3(CO)_{12}$, 1953 cm^{-1} (○) associated with $Ru(CO)_4PPh_3$, and 1910 cm^{-1} (×) associated with $Ru(CO)_3(PPh_3)_2$. Spectral changes are for 454.4-nm irradiation of 3.3×10^{-4}M $Ru_3(CO)_{12}$, 0.082M PPh_3 in isooctane under nitrogen.

tene yields some conversion to pentane. The $Os_3(CO)_{10}H_2$ has been known for some time (93) and is a thermal reaction product of $Os_3(CO)_{12}$ and H_2 (94).

The disappearance quantum yields for the $M_3(CO)_{12}$ complexes are given in Table XI. Quite interestingly, the quantum yields are very low; even in neat solutions of alkene or $HSiEt_3$ the yields are small.

Discussion

The data adequately support the conclusion that $M_3(CO)_{12}$ (M = Fe, Ru, Os) are effective catalyst precursors when irradiated with wavelengths corresponding to electronic excitation. In terms of both quantum

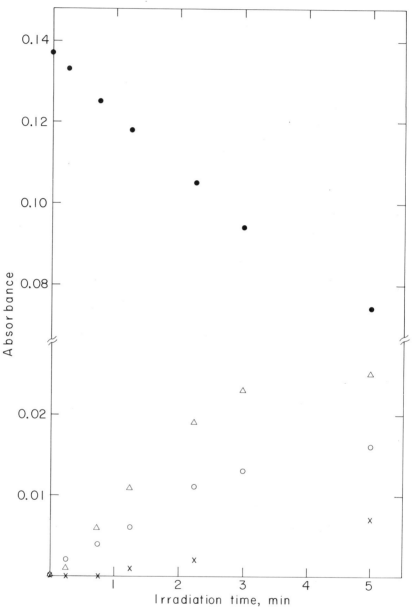

Figure 4. Plots of absorbance against 355-nm irradiation time for ~10^{-4}M $Os_3(CO)_{12}$, 0.01M PPh_3 in isooctane under nitrogen. Bands monitored are 2070 cm^{-1} (●) $Os_3(CO)_{12}$; 2020 cm^{-1} (△) and 2055 cm^{-1} (○), both $Os_3(CO)_{11}PPh_3$; and 1979 cm^{-1} (×) $Os_3(CO)_9(PPh_3)_3$.

Table XI. $M_3(CO)_{12}$ Disappearance Quantum Yields[a]

M	Irrdn. λ (nm)	Solvent	Φ ± 20%
Fe	355	isooctane	0.007
		1-pentene	0.01
		HSiEt$_3$	0.04
		1-pentene:HSiEt$_3$ (1:1)	0.04
	633	0.001M PPh$_3$ in isooctane	0.01
		0.09M PPh$_3$ in isooctane	0.02
Ru	355	isooctane	< 10^{-3}
		1-pentene	0.03
		HSiEt$_3$	0.03
		1-pentene:HSiEt$_3$ (1:1)	0.03
Os	355	1.0M 1-pentene in isooctane	0.03
		1.0M 1-pentene:1.0M HSiEt$_3$ in isooctane	0.02
		HSiEt$_3$	0.02

[a] All data are for degassed solutions at 298°K.

yield (>> unity) and the number of molecules reacted per $M_3(CO)_{12}$ initially present, we can state that a true catalyst is generated from the photolysis of $M_3(CO)_{12}$. Generally, neat solutions of a terminal alkene and a silicon hydride can be converted to products to an extent exceeding 90% with only ~ 10^{-3}M $M_3(CO)_{12}$. Such reactants are typically consumed with initial quantum yields that exceed 10 while the $M_3(CO)_{12}$ disappears with an initial quantum yield of < 0.1. If the catalytic intermediates do not regenerate $M_3(CO)_{12}$, we can assign very high turnover numbers to the intermediates. Since the actual catalytically active species very likely involves the loss of at least one CO molecule (95), any regeneration of $M_3(CO)_{12}$ is likely to be very slow on the time scale of the quantum yield determination. It is reasonable that the regeneration of $M_3(CO)_{12}$ would be slow in solutions that contain as much alkene and silicon hydride as used here.

Turning to the catalytic chemistry itself, we note that there are a large number of catalysts for the hydrosilation of olefins (95). However, there seem to be few catalysts that effect the generation of alkenyl silanes along with the alkyl silanes. Indeed, as far as homogeneous systems are concerned, it appears that only Fe(CO)$_5$ is an effective catalyst for the formation of vinylsilanes from an alkene and a silicon hydride. Previously, we reported (41) that Fe(CO)$_5$ is an effective photocatalyst for this reaction, and the data herein show that the clusters $M_3(CO)_{12}$ all give some vinylsilane product. The (pentenyl)SiEt$_3$ products are least important for M = Os but for both M = Fe and Ru, the (pentenyl)SiEt$_3$

product is the major (under some conditions the exclusive) silicon-containing catalysis product. The synthesis of such products is not particularly unusual since hydrosilation of the appropriate alkyne can lead to the vinylsilane. Vinylsilanes having no acetylenic precursor can be synthesized (*41*) by the photocatalysis procedure, but the rate of the photocatalyzed formation of vinylsilanes from cycloalkenes or 1,1-disubstituted alkenes is likely to be much slower. This follows from our observation that $M_3(CO)_{12}$ selectively reacts with terminal alkenes. As has been found before, the distinct advantage of the photocatalysis procedure is that the temperature of the reaction is very low compared with the typical catalysis conditions.

The fact that the $M_3(CO)_{12}$ species selectively catalyzes reaction of the terminal alkene and, by comparison, does not equilibrate alkenes by isomerization at a fast rate can be advantageous. For example, the reaction indicated in 16 seems doable by the $Ru_3(CO)_{12}$ photocatalysis

$$\text{cyclopentene} \xrightarrow{HSiR_3} \text{cyclopentene-SiR}_3 + \text{cyclopentane} \qquad (16)$$

procedure, owing to the fact that Reactions 17 and 18 and subsequent

$$\text{cyclopentene} \longrightarrow \text{cyclopentene} \qquad (17)$$

$$\text{cyclopentadiene} \longrightarrow \text{cyclopentene-SiR}_3 + \text{cyclopentane} \qquad (18)$$

isomerization of the vinylsilane are likely to be slow. Further, the vinylsilane product in Reaction 16 has no acetylenic precursor. A report on attempts to produce such products in synthetic quantities by the photocatalysis procedure will be the object of a future paper. The lack of isomerization activity on the same scale as the reaction with the silicon hydride can be a disadvantage, too. Such is the case in the attempted reaction of a mixture of the pentenes to yield (pentyl)$SiEt_3$ and (pentenyl)$SiEt_3$. Slow equilibration of the pentenes allows consumption of the 1-pentene, and production of more 1-pentene is the rate-limiting process.

At this stage, the details of the mechanism for photocatalytic activity of $M_3(CO)_{12}$ are not elucidated completely, but some key facts are certain. For the iron and ruthenium clusters, the organometallic photoproducts that result are mononuclear species. Although the disappearance yields are low, irradiation of $M_3(CO)_{12}$ (M = Fe, Ru) gives good chemical yields of mononuclear products when irradiation is carried out in the presence of alkene, CO, or PPh_3. Further, in the presence of PPh_3, we find that $M(CO)_3(PPh_3)_2$ (M = Fe, Ru) is the primary photoproduct. Thus, for the iron and ruthenium clusters, we propose that fragmentation to yield catalytically active mononuclear species is the result of irradiation in the presence of 1-pentene and $HSiEt_3$. Indeed, for M = Fe, the cluster gives nearly the same distribution of silicon-containing products as when using $Fe(CO)_5$ as the catalyst precursor. We suggest that the catalytically repeating unit is "$M(CO)_3$" (M = Fe, Ru), as shown in Reactions 19–29. The catalytic cycle is illustrated for alkene = propene. Naturally,

$$\text{"M(CO)}_3\text{"} \xrightarrow{} M(CO)_3(\diagup\!\!\!\diagdown) \tag{19}$$

$$\text{"M(CO)}_3\text{"} \xrightarrow{HSiR_3} HM(CO)_3(SiR_3) \tag{20}$$

$$M(CO)_3(\diagup\!\!\!\diagdown) \rightleftarrows HM(CO)_3(\eta^3\text{-}C_3H_5) \tag{21}$$

$$M(CO)_3(\diagup\!\!\!\diagdown) \xrightarrow{HSiR_3} HM(CO)_3(\diagup\!\!\!\diagdown)(SiR_3) \tag{22}$$

$$HM(CO)_3(SiR_3) \xrightarrow{} HM(CO)_3(\diagup\!\!\!\diagdown)(SiR_3) \tag{23}$$

$$HM(CO)_3(\diagup\!\!\!\diagdown)(SiR_3) \longrightarrow H\text{—}M(CO)_3\text{—}SiR_3 \tag{24}$$

$$(25)$$

$$(26)$$

$$(27)$$

$$H\text{—}M(CO)_3 \longrightarrow \text{"M(CO)}_3\text{"} + \diagup\!\!\!\diagdown \tag{28}$$

$$\longrightarrow \text{"M(CO)}_3\text{"} + \diagup\!\!\!\diagdown\text{SiR}_3 \tag{29}$$

"M(CO)$_3$" likely does not exist as such since the reactions typically are carried out in the presence of very high concentrations of alkene and silane. This is the mechanism previously proposed for Fe(CO)$_5$ (*30, 41, 86*). Organic or silicon radicals are not likely too important in the catalysis since each metal cluster gives a different distribution of organosilane products. Two practical advantages can be associated with using the Fe$_3$(CO)$_{12}$ compared with Fe(CO)$_5$. First, Fe$_3$(CO)$_{12}$ is a solid that is not too volatile and that can be handled more conveniently than Fe(CO)$_5$. Second, the Fe$_3$(CO)$_{12}$ absorbs, and is effective, throughout the visible spectrum, whereas Fe(CO)$_5$ is virtually colorless at $\sim 10^{-3} M$ and requires UV excitation. There is a tendency, though, for the Fe$_3$(CO)$_{12}$ photocatalysts to give smaller-extent conversions upon visible excitation. This is likely caused by Reactions like 30 and 31 which may

$$\text{HM(CO)}_3(\text{SiR}_3) \xrightarrow{\text{CO}} \text{HM(CO)}_4(\text{SiR}_3) \qquad (30)$$

$$\text{M(CO)}_3(\rlap{=}\diagup) \xrightarrow{\text{CO}} \text{M(CO)}_4(\rlap{=}\diagup) \qquad (31)$$

occur. The CO could result from some decomposition of metal carbonyl species. To generate the "M(CO)$_3$" species from the tetracarbonyl products requires UV excitation.

The Os$_3$(CO)$_{12}$-photocatalyzed reactions and its own photoproducts are qualitatively different compared with iron and ruthenium clusters. On the basis of the persistance of Os$_3$ cluster products, it is very tempting to conclude that the photocatalysis involves clusters as the catalytically active species. While this point needs further deliberation, it is gratifying to note the clean generation of Os$_3$(CO)$_{10}$H$_2$ from irradiation of Os$_3$(CO)$_{12}$ under hydrogen. This fact, along with the observation of photocatalyzed 1-pentene hydrogenation, suggest a photoacceleration of the known Os$_3$(CO)$_{10}$H$_2$ hydrogenation of alkenes (*71, 72*). A role for Os$_3$ units in the photocatalysis is implicated and will be the object of future studies.

One final point merits discussion. The primary photoprocess in M$_3$(CO)$_{12}$ is likely the cleavage of one of the metal–metal bonds to form the diradical indicated in Reaction 9. The low disappearance quantum yields, compared with the declusterification of dinuclear metal–metal bonded species (*9*), are explicable in terms of efficient closure (reverse of Reaction 9) to regenerate the cluster. The essential independence of the quantum yields on substrate concentration suggests that the diradical undergoes some fast, unimolecular decomposition, perhaps as in Reaction 32, for M = Fe, Ru. The dinuclear, formally M–M double bonded,

$$(\text{OC})_4\text{M}\diagdown_{\text{M(CO)}_4}\diagup\text{M(CO)}_4 \xrightarrow{\Delta} \text{M}_2(\text{CO})_8 + \text{M(CO)}_4 \qquad (32)$$

$M_2(CO)_8$ and the $M(CO)_4$ can both react with CO to give ultimately $M(CO)_5$. From studies with $Fe(CO)_5$, (30), it is known that a primary isolable photoproduct is the disubstituted $Fe(CO)_3(PPh_3)_2$ when the irradiation is carried out in the presence of PPh_3. Thus, $Fe(CO)_4$ at least is a viable precursor to "$M(CO)_3$" catalytic species. The primary photoprocess in $Os_3(CO)_{12}$ is very likely Os–Os bond cleavage as well, since mononuclear products have been observed after prolonged irradiation, e.g., Ref. 59. The interaction with hydrogen may occur as in Reaction 33. Triple-bonded complexes have been observed to be photo-

$$(OC)_4Os\diagup^{Os(CO)_4}\diagdown_{Os(CO)_4} \xrightarrow{-2CO} (OC)_3Os\diagup^{Os(CO)_4}\diagdown_{Os(CO)_3}\!\!\!\!\!\!\!\equiv\!\!\!\!\!\!\!Os(CO)_3 \xrightarrow{H_2} Os(CO)_{10}(H_2) \quad (33)$$

chemically generated from single-bonded complexes (81), presumably via a similar mechanism involving labile, metal-centered radicals.

Experimental

Materials. All solvents, substrates, and catalyst precursors were obtained commercially. Isooctane was spectroquality, and the alkenes were the purest materials obtainable from Chemical Samples Co. The alkenes typically were passed through alumina immediately prior to use to remove peroxides. The $HSiEt_3$ was distilled prior to use, and the purity of $M_3(CO)_{12}$ complexes was determined by IR after sublimation. $Fe(CO)_5$ was used after distillation. Authentic samples of the (pentyl)-$SiEt_3$ and (pentenyl)$SiEt_3$ were prepared and characterized as for (pentyl)$SiMe_3$ and (pentenyl)$SiMe_3$ (41).

Spectra. All IR spectra were recorded using a Perkin–Elmer 180 spectrometer with 0.1- or 1.0-mm matched pathlength cells. Photoproducts identified by IR were compared with literature data (see Table X) and identified by independent generation from $Fe(CO)_5$ for products from $Fe_3(CO)_{12}$. Appropriate care was taken in the handling of air-sensitive species. UV-visible spectra were recorded using a Cary 17.

Irradiation Sources. Several different sources were used in this work. For 633-nm irradiations, a 6X beam expanded He–Ne laser ($\sim$ 5 mW output) was used. The 454.4-nm irradiation was from an argon ion laser (Spectra Physics Model 164). Irradiations at 550 nm were carried out using a 550 W, medium-pressure Hg lamp from Hanovia, filtered with the appropriate Corning glass filter pack to isolate the 550-nm emission. Most irradiations were carried out using a GE Black Light equipped with two 15-W fluorescent Black Light bulbs. The output of the lamp is centered at 355 nm, and the width at half-height is $\sim$ 25 nm. The intensity of the light was determined using ferrioxalate actinometry (96), and the typical dose for the samples was 2×10^{-6} ein/min.

Irradiation Procedures. 1.0, 2.0, or 3.0 mL samples of alkene and/or $HSiEt_3$ with $10^{-3} M$ $M_3(CO)_{12}$ or $Fe(CO)_5$ were put in 13 x 100-mm borosilicate test tubes with constrictions. The samples were freeze-pump-thaw degassed to 10^{-5} Torr in at least three cycles and were hermetically sealed. The samples then were irradiated for given periods of time at 298°K and analyzed.

Analysis. Alkene isomerization and reaction with silane was monitored by GC under the conditions previously reported (30, 41).

Acknowledgment

We thank the Office of Naval Research for support of this research, and MSW acknowledges support as a Dreyfus Teacher-Scholar Grant Recipient, 1975–1980.

Literature Cited

1. Wrighton, M., *Chem. Rev.* (1974) **74**, 401.
2. Wrighton, M. S., *Top. Curr. Chem.* (1976) **65**, 37.
3. Balzani, V., Carassiti, V., "Photochemistry of Coordination Compounds," Academic, New York, 1970.
4. "Concepts of Inorganic Photochemistry," A. W. Adamson, P. D. Fleischauer, Eds., John Wiley, New York, 1975.
5. Koerner von Gustorf, E., Grevels, F.-W., *Top. Curr. Chem.* (1969) **13**, 366.
6. Malouf, G., Ford, P. C., *J. Am. Chem. Soc.* (1974) **96**, 601.
7. Wrighton, M. S., Abrahamson, H. B., Morse, D. L., *J. Am. Chem. Soc.* (1976) **98**, 4105.
8. Giordano, P. J., Wrighton, M. S., *Inorg. Chem.* (1977) **16**, 160.
9. Wrighton, M. S., Ginley, D. S., *J. Am. Chem Soc.* (1975) **97**, 2065, 4246.
10. Byers, B. H., Brown, T. L., *J. Am. Chem. Soc.* (1975) **97**, 3270, 947.
11. Hughey, J. L., Bock, C. R., Meyer, T. J., *J. Am. Chem. Soc.* (1975) **97**, 4440.
12. Giannotti, C., Merle, G., *J. Organomet. Chem.* (1976) **105**, 97.
13. Laine, R. M., Ford, P. C., *Inorg. Chem.* (1977) **16**, 388.
14. Hudson, A., Lappert, M. F., Nicholson, B. K., *J. Chem. Soc., Dalton Trans.* (1977) 551.
15. Abrahamson, H. B., Wrighton, M. S., *J. Am. Chem. Soc.* (1977) **99**, 5510.
16. Geoffroy, G. L., Gray, H. B., Hammond, G. S., *J. Am. Chem Soc.* (1975) **97**, 3933.
17. Geoffroy, G. L., Bradley, M. G., *Inorg. Chem.* (1977) **16**, 744.
18. Geoffroy, G. L., Pierantozzi, P., *J. Am. Chem. Soc.* (1976) **98**, 8054.
19. Collman, J. P., *Acc. Chem. Res.* (1968) **1**, 136.
20. Halpern, J., *Acc. Chem. Res.* (1970) **3**, 368.
21. Cotton, F. A., Wilkinson, G., "Advanced Inorganic Chemistry," 3rd ed., Chap. 24, pp. 770–801, Interscience, New York, 1972.
22. Wrighton, M., Ginley, D. S., Schroeder, M. A., Morse, D. L., *Pure Appl. Chem.* (1975) **41**, 671.
23. Frankel, E. N., Emken, E. A., Peters, H. M., Davison, V. L., Butterfield, R. O., *J. Org. Chem.* (1964) **29**, 3292, 3299.
24. Frankel, E. N., Emken, E. A., Davison, V. L., *J. Org. Chem.* (1965) **30**, 2739.
25. Cais, M., Moaz, N., *J. Chem. Soc. A* (1971) 1811.
26. Moaz, N., Cais, M., *Isr. J. Chem.* (1968) **6**, 32.
27. Ogata, I., Misono, A., *J .Chem. Soc. J.* (1974) **85**, 748, 853.
28. Poliakoff, M., Turner, J. J., *J. Chem. Soc., Dalton Trans.* (1974) 2276.
29. Poliakoff, M., *J. Chem. Soc., Dalton Trans.* (1974) 210.
30. Schroeder, M. A., Wrighton, M. S., *J. Am. Chem. Soc.* (1976) **98**, 551.

31. Wrighton, M., Hammond, G. S., Gray, H. B., *J. Organomet. Chem.* (1974) **70**, 283.
32. Wrighton, M., Hammond, G. S., Gray, H. B., *J. Am. Chem. Soc.* (1970) **92**, 6068.
33. Nasielski, J., Kirsch, P., Wilputte-Steinert, L., *J. Organomet. Chem.* (1971) **27**, C13.
34. Wrighton, M., Schroeder, M. A., *J. Am. Chem. Soc.* (1973) **95**, 5764.
35. Platbrood, G., Wilputte-Steinert, L., *J. Organomet. Chem.* (1974) **70**, 393, 407.
36. Platbrood, G., Wilputte-Steinert, L., *J. Organomet. Chem.* (1975) **85**, 199.
37. Platbrood, G., Wilputte-Steinert, L., *Tetrahedron Lett.* (1974) 2507.
38. Rietvelde, D., Wilputte-Steinert, L., *J. Organomet. Chem.* (1976) **118**, 191.
39. Fischler, I., Budzwait, M., Koerner von Gustorf, E. A., *J. Organomet. Chem.* (1976) **105**, 325.
40. Wrighton, M. S., Schroeder, M. A., *J. Am. Chem. Soc.* (1974) **96**, 6235.
41. Schroeder, M. A., Wrighton, M. S., *J. Organomet. Chem.* (1977) **128**, 345.
42. Jennings, W., Hill, B., *J. Am. Chem. Soc.* (1970) **92**, 3199.
43. Hill, B., Math, K., Pillsbury, D., Voecks, G., Jennings, W., *Mol. Photochem.* (1973) **5**, 195.
44. Krausz, P., Garnier, F., Dubois, J. E., *J. Am. Chem. Soc.* (1975) **97**, 437.
45. Agapiou, A., McNelis, E., *J. Chem. Soc., Chem Commun.* (1975) 187.
46. Agapiou, A., McNelis, E., *J. Organomet. Chem.* (1975) **99**, C47.
47. Giannotti, C., Green, M. L. H., *J. Chem. Soc., Chem. Commun.* (1972) 1114.
48. Tang Wong, K. L., Thomas, J. L., Brintzinger, H. H., *J. Am. Chem. Soc.* (1974) **96**, 3694.
49. Johnson, B. F. G., Lewis, J., Twigg, M. V., *J. Organomet. Chem.* (1974) **67**, C75.
50. Johnson, B. F. G., Lewis, J., Twigg, M. V., *J. Chem. Soc., Dalton Trans.* (1976) 1876.
51. Knox, S. A. R., Stone, F. G. A., *J. Chem. Soc. A* (1971) 2874.
52. Knox, S. A. R., Stone, F. G. A., *J. Chem. Soc. A* (1970) 3147.
53. Knox, S. A. R., Stone, F. G. A., *J. Chem. Soc. A* (1969) 2559.
54. Brockes, A., Knox, S. A. R., Stone, F. G. A., *J. Chem. Soc. A* (1971) 3469.
55. Cullen, W. R., Harbourne, D. A., *Inorg. Chem.* (1970) **9**, 1839.
56. Cullen, W. R., Harbourne, D. A., Liengme, B. V., Sams, J. R., *Inorg. Chem.* (1970) **9**, 702.
57. Roberts, P. J., Trotter, J., *J. Chem. Soc. A* (1971) 1479.
58. Cullen, W. R., Harbourne, D. A., Liengme, B. V., Sams, J. R., *J. Am. Chem. Soc.* (1968) **90**, 3293.
59. Cotton, F. A., Deeming, A. J., Josty, P. L., Ullah, S. S., Domingos, A. J. P., Johnson, B. F. G., Lewis, J., *J. Am. Chem. Soc.* (1971) **93**, 4624.
60. Asinger, F., Fell, B., Schrage, K., *Chem. Ber.* (1965) **98**, 372.
61. Manuel, T. A., *J. Org. Chem.* (1962) **27**, 3941.
62. Carr, M. D., Kane, V. V., Whiting, M. C., *Proc. Chem. Soc.* (1964) 408.
63. Casey, C. P., Cyr, C. R., *J. Am. Chem. Soc.* (1973) **95**, 2248.
64. Alper, H., LePort, P. C., *J. Am. Chem. Soc.* (1969) **91**, 7553.
65. Bingham, D., Hudson, B., Webster, D. E., Wells, P. B., *J. Chem. Soc., Dalton Trans.* (1974) 1521.
66. Castiglioni, M., Milone, L., Ostella, D., Vaglio, G. A., Valle, M., *Inorg. Chem.* (1976) **15**, 394.
67. Canty, A. J., Johnson, B. F. G., Lewis, J., *J. Organomet. Chem.* (1972) **43**, C35.
68. Canty, A. J., Domingos, A. J. P., Johnson, B. F. G., Lewis, J., *J. Chem. Soc., Dalton Trans.* (1973) 2056.

69. Deeming, M., Underhill, M., *J. Chem. Soc., Dalton Trans.* (1974) 1415, and references therein.
70. Deeming, A. J., Hasso, S., Underhill, M., *J. Chem. Soc., Dalton Trans.* (1975) 1614, and references therein.
71. Keister, J. B., Shapley, J. R., *J. Organomet. Chem.* (1975) **85**, C29.
72. Keister, J. B., Shapley, J. R., *J. Am. Chem. Soc.* (1976) **98**, 1056.
73. Calvert, R. B., Shapley, J. R., *J. Am. Chem. Soc.* (1977) **99**, 5525, and references therein.
74. McCleverty, J. A., *J. Organomet. Chem.* (1975) **89**, 273.
75. McCleverty, J. A., *J. Organomet. Chem.* (1976) **119**, 261.
76. Ferrari, R. P., Vaglio, G. A., *Gazz. Chim. Ital.* (1975) **105**, 939.
77. Gambino, O., Valle, M., Aime, S., Vaglio, G. A., *Inorg. Chim. Acta* (1974) **8**, 71.
78. Gambino, O., Ferrari, R. P., Chinone, M., Vaglio, G. A., *Inorg. Chim. Acta* (1975) **12**, 155.
79. Tachikawa, M., Shapley, J. R., *J. Organomet. Chem.* (1977) **124**, C19.
80. Bond, G. C., Hellier, M., *J. Catal.* (1965) **4**, 1.
81. Ginley, D. S., Bock, C. R., Wrighton, M. S., *Inorg. Chim. Acta* (1977) **23**, 85.
82. Byers, B. H., Brown, T. L., *J. Am. Chem. Soc.* (1977) **99**, 2527.
83. Absi-Halalbi, M., Brown, T. L., *J. Am. Chem. Soc.* (1977) **99**, 2982.
84. Angelici, R. J., Siefert, E. E., *Inorg. Chem.* (1966) **5**, 1457.
85. Candlin, J. P., Shortland, A. C., *J. Organomet. Chem.* (1969) **16**, 289.
86. Bruce, M. I., Stone, F. G. A., *Angew. Chem. Inst. Ed. Eng.* (1968) **7**, 427.
87. Poë, A., Twigg, M. V., *J. Chem. Soc., Dalton Trans.* (1974) 1860.
88. Poë, A., Twigg, M. V., *Inorg. Chem.* (1974) **13**, 2982.
89. Bruce, M. I., Shaw, G., Stone, F. G. A., *J. Chem. Soc., Dalton Trans.* (1972) 2094.
90. Bradford, C. W., van Bronswijk, W., Clark, R. J. H., Nyholm, R. S., *J. Chem. Soc., Dalton Trans.* (1970) 2889.
91. Bradford, C. W., Nyholm, R. S., *Chem. Commun.* (1967) 384.
92. L'Eplattenier, F., Calderazzo, F., *Inorg. Chem.* (1968) **7**, 1290.
93. Johnson, B. F. G., Lewis, J., Kilty, P. A., *J. Chem. Soc. A* (1968) 2859.
94. Kaesz, H. D., Knox, S. A. R., Koepke, J. W., Saillant, R. B., *Chem. Commun.* (1971) 477.
95. Harrod, J. F., Chalk, A. J., "Organic Synthesis via Metal Carbonyls," Vol. 2, I. Wender, P. Pino, Eds., pp. 673–704, John Wiley, New York, 1977.
96. Hatchard, C. G., Parker, C. A., *Proc. R. Soc. London, A* (1956) **235**, 518.
97. Knight, J., Mays, M. J., *Chem. Commun.* (1970) 1006.
98. Murdoch, H. D., Weiss, E., *Helv. Chim. Acta* (1963) **46**, 1588.
99. Calderazzo, F., L'Eplattenier, F., *Inorg. Chem.* (1967) **6**, 1220.
100. Johnson, B. F. G., Johnston, R. D., Josty, P. L., Lewis, J., Williams, I. G., *Nature* (1967) **213**, 901.

RECEIVED September 20, 1977.

AUTHOR INDEX

Names and numbers in bold are authors and first pages, respectively, of complete chapters. Numbers in parentheses are reference numbers which indicate that an author's work is referred to, followed by pages of that referral. Numbers in italics show the page on which the complete reference is listed.

A

Abrahamson, H. B.   (37) 139, 141, *145;* (7) 190, *212;* (15) 190–192, *212*
Abruna, H.   28; (2) 73, 89
Absi-Halalbi, M.   (83) 201, *214*
Adamson, A. W.   (1) 1, *25;* (2) 60, *71;* (5) 73, *89;* (6) 73, 77, *89;* (28) 84, *90;* (20) 104, *114;* (40) 169, 171, *173;* (4) 189, *212*
Agapiou, A.   (45) 191, *213;* (46) 191, *213*
Aime, S.   (77) 192, *214*
Albano, V. G.   (3) 132, *145;* (29) 135, *145*
Albertin, G.   (4) 46, *56*
Ali, L. H.   (23) 116, *131*
Allinger, N. L.   (7) 161, *172*
Allsopp, S. R.   (8) 65, *72*
Alper, H.   (64) 192, *213*
Alway, D. G.   115; (26) 117, 124, 128, *131;* (27) 117, 121, *131*
Anbar, M.   (27) 81, *90*
Anderson, A. S.   (42) 129, *131*
Anderson, C. B.   (48) 185, *187*
Anderson, C. P.   (18) 2, 17, *26;* (12) 116, 121, *130*
Angelici, R. J.   (84) 203, *214*
Aoyagui, S.   (16) 2, 7, *26;* (33) 10, *26*
Archer, L. J.   (13) 148, *157*
Aresta, M.   (12) 148, *157*
Arnold, D. R.   (1) 174, 179, *186*
Asinger, F.   (60) 192, *213*
Atwood, J. D.   (40) 127, *131*
Atwood, J. L.   (36) 166, *173*
Austin, R. G.   189

B

Baker, B. R.   (45) 20, *26;* (15) 94, *113*
Baldwin, J. E.   (9) 174, 179, *187*
Ballardini, R.   (52) 20, *27;* (7) 92, *113;* (19) 103, *114;* (41) 169, 171, *173*
Balzani, V.   (4) 1, *25;* (5) 1, *25;* (22) 3, *26;* (52) 20, *27;* (1) 60, *71;* (10) 66, *72;* (3) 91–95, 100, 103, 104, 108, 109, *113;* (5) 92, *113;* (7) 92, *113;* (9) 93, *113;* (12) 93, 111, *113;* (21) 104, *114;* (3) 115, *130;* (41) 169, 171, *173;* (3) 189, *212*
Bard, A. J.   (15) 2, 5, 7, 23, *26*
Barnett, K. W.   115; (17) 116, 120, *131;* (22) 116, *131;* (26) 117, 124, 128, *131;* (27) 117, 121, *131;* (30) 117, 125, 126, *131;* (31) 117, 125, 126, *131;* (39) 127, *131*
Basolo, F.   (11) 66, *72;* (35) 121, *131*
Baxendale, J. H.   (14) 2, *26*
Bayless, J.   (39) 183, *187*
Beach, D. L.   (30) 117, 125, 126, *131;* (31) 117, 125, 126, *131;* (39) 127, *131*
Behrendt, S.   (28) 107, *114*
Beier, B. F.   (30) 108, 111, *114*
Bensasson, R.   (22) 3, *26*
Betkouski, M.   (12) 174, 179, *187*
Biedermann, G.   (37) 12, *26*
Bingham, D.   (65) 192, *213*
Bishop, K. C.   (5) 160, 172, *172*
Bock, C. R.   (2) 1, *25;* (26) 3, 5, 7, *26;* (21) 76, 77, *90;* (11) 116, 121, *130;* (18) 134, *145;* (11) 190–192, *212;* (81) 201, 211, *214*
Bolletta, F.   (5) 1, *25;* (10) 66, *72;* (3) 91–95, 100, 103, 104, 108, 109, *113;* (5) 92, *113;* (9) 93, *113;* (12) 93, 111, *113*
Bommer, J. C.   (35) 166, *173*
Bond, G. C.   (80) 195, *214*
Bor, G.   (33) 138, *145*
Borčić, S.   (36) 182, 183, *187*
Böttcher, W.   (32) 7, 10, 15, 20, 22, *26*
Braddock, J. N.   (12) 67, *72*
Bradford, C. W.   (90) 203, *214;* (91) 203, *214*
Bradley, M. G.   (17) 190, *212*
Brant, P.   (24) 154, *157*

215

Bremner, J. B.   (8) 175, 179, *187*
Breslow, R.   (46) 143, 144, *146;* (47) 143, 144, *146;* (30) 182, 183, *187*
Brinton, R. K.   (46) 183, *187*
Brintzinger, H. H.   (48) 191, *213*
Brockes, A.   (54) 192, 201, *213*
Brookes, A.   (13) 132, 133, *145*
Brown, G.   (47) 20, *26*
Brown, G. M.   (57) *27*
Brown, T. L.   (14) 116, 121, *130;* (40) 127, *131;* (49) 144, *146;* (10) 190–192, 201, *212;* (82) 201, *214;* (83) 201, *214*
Brubaker, C. H.   (44) 130, *131*
Bruce, M. I.   (9) 132, 133, *145;* (30) 165, *173;* (31) 165, *173;* (32) 165, *173;* (86) 203, 210, *214;* (89) 203, *214*
Brunschwig, B.   (50) 18, *27*
Budzwait, M.   (39) *213*
Burdett, J. K.   (5) 115, *130*
Burreson, B. J.   (48) 185, *187*
Busby, D. C.   **147;** (17) 149, *157;* (25) 154, 156, *157*
Butler, J. R.   (45) 183, *187*
Butterfield, R. O.   (23) 191, *212*
Byers, B. H.   (14) 116, 121, *130;* (49) 144, *146;* (10) 190–192, 201, *212;* (82) 201, *214*

C

Cable, J. W.   (36) 139, *145*
Cais, M.   (25) 191, *212;* (26) 191, *212*
Calderazzo, F.   (92) 203, 204, *214;* (99) 203, *214*
Calvert, J. G.   (33) 117, *131*
Calvert, R. B.   (73) 192, *214*
Candlin, J. P.   (54) 13, 22, *27;* (85) 203, 205, *214*
Canty, A. J.   (67) 192, *213;* (68) 192, *213*
Carassiti, V.   (1) 60, *71;* (3) 115, *130;* (3) 189, *212*
Cargill, R. L.   (9) 161, *172*
Carlyle, D. W.   (56) 22, 23, *27*
Carr, M. D.   (62) 192, *213*
Casagrande, G. T.   (25) 135, *145*
Casey, C. P.   (63) 192, *213*
Cassadonte, D.   (22) 179, 182, *187*
Cassar, L.   (49) 185, *187*
Castelli, F.   (11) 93, *113*
Castiglioni, M.   (66) 192, *213*
Chaisson, D. A.   (8) 73, *89;* (9) 73, 74, 77, 82, *89*
Chalk, A. J.   (95) 20, *214*
Chang, M.   (20) 149, *157*
Chatt, J.   (2) 147, *156;* (14) 174, *187;* (16) 148, 149, *157;* (18) 149, *157*
Chini, P.   (2) 132, *145;* (3) 132, *145;* (27) 135, *145;* (29) 135, *145*
Chinone, M.   (78) 192, *214*
Chou, M.   (32) 7, 10, 15, 20, 22, *26;* (53) 13, 20, 22, *27*

Churchill, M. R.   (32) 165, *173*
Cimolino, M.   (4) 46, *56*
Clark, R. J. H.   (90) 203, *214*
Clark, W. D. K.   (12) 1, 6, *26;* (57) *27*
Colli, L.   (27) 135, *145*
Collman, J. P.   (19) 190, *212*
Connon, H. A.   (4) 160, *172*
Conway, J.   (2) 91, 94, 96, 107, *113*
Cooke, M.   (9) 132, 133, *145;* (28) 135, *145*
Cope, A. C.   (11) 174, 179, *187;* (23) 164, *173*
Cotton, F. A.   (8) 132, 133, *145;* (20) 164, *173;* (21) 190, *212;* (59) 192, 201, 211, *213*
Cox, A.   (8) 65, 72; (23) 116, *131*
Creutz, C.   **1;** (9) 1, 21, *26;* (17) 2, 15, 17, *26;* (27) 5, 7, 15, 18, *26;* (32) 7, 10, 15, 20, 22, *26;* (41) 15, *26;* (42) 7, 15–17, 22, *26;* (44) 22, 23, 25, *26;* (53) 13, 20, 22, *27;* (4) 73, *89;* (23) 77, *90;* (43) 170, *173*
Crosby, G. A.   (20) 2, *26;* (24) 3, *26;* (31) 9, *26;* (14) 67, *72;* (15) 67, *72;* (16) 68, *72;* (31) *90;* (39) 171, *173*
Cullen, W. R.   (14) 132, 134, *145;* (15) 132, 134, *145;* (16) 132, 134, *145;* (55) 192, 201, *213;* (56) 192, 201, *213;* (58) 192, 201, *213*
Cyr, C. R.   (63) 192, *213*

D

Dahl, L. F.   (22) 135, *145;* (41) 142, *146*
Daniels, F.   (3) 159, *172*
Dannöhl-Fickler, R.   (22) 105, *114*
Darensbourg, D. J.   (41) 127, 128, *131;* (9) 148, *157*
Dattilo, M.   (31) 117, 125, 126, *131*
Dauben, W. G.   (9) 161, *172*
Davison, V. L.   (23) 191, *212;* (24) 191, *212*
Day, V. W.   (3) 147, 154, *156;* (4) 148, 149, 154, *156;* (5) 147, 154, *156*
DeBoer, B. G.   (32) 165, *173*
Deboer, C. D.   (12) 161, *173*
Deeming, A. J.   (59) 192, 201, 211, *213;* (70) 192, *214*
Deeming, M.   (59) 192, *214*
Delaive, P. J.   **28;** (2) 73, *89*
Dellinger, B.   (18) 69, *72*
Demas, J. N.   (8) 1, 8, *26;* (20) 2, *26;* (4) 61, *71;* (5) 61, *71;* (5) 73, *89;* (40) 169, 171, *173*
Deno, N. C.   (31) 182, 183, *187*
Dent, W. T.   (26) 135, *145*
DePuy, C. H.   (28) 182, *187;* (29) 182, *187*
Dewar, M. J. S.   (32) 182, 183, *187*
Deyrup, J. A.   (12) 174, 179, *187*
Diamantis, A. A.   (2) 147, *156;* (18) 149, *157*
Dobson, G. R.   (4) 115, *130*

Dockal, E. R. (61) 13, 27
Doepker, R. D. (47) 183, *187*
Domingos, A. J. P. (59) 192, 201, 211, *213*; (68) 192, *213*
Doyle, J. (59) 13, 27
DuBios, D. L. (22) 152, 154, 156, *157*
Dubois, J. E. (44) 191, *213*
Duncanson, L. A. (26) 135, *145*
Durante, V. A. (14) 73, 77, 87, *89;* (15) 73, 5, 82, 88, *89;* (17) 73, 88, *89*

E

Eaton, P. E. (49) 185, *187*
El Sandi, N. (17) 175, *187*
Emken, E. A. (23) 191, *212;* (24) 191, *212*
Endicott, J. F. (17) 95, *114;* (62) 13, 27
Enos, C. T. (38) 141, *145*
Epstein, R. A. 132
Ercoli, R. (25) 135, *145*
Erwin, D. K. (11) 53, *56*
Espenson, J. H. (56) 22, 23, 27

F

Faller, J. W. (42) 129, *131*
Faraggi, M. (55) 22, 27; (63) 13, 27
Feder, A. (55) 22, 27; (63) 13, 27
Fell, B. (60) 192, *213*
Feltham, R. D. (24) 154, *157*
Fenske, R. F. (25) 117, 123, 126, *131*
Ferguson, J. (18) 96, *114*
Ferguson, J.A. (36) 121, *131*
Ferrari, R. P. (76) 192, *214;* (78) 192, *214*
Ferraro, J. R. (24) 105, *114*
Fieldhouse, S. A. (30) 135, *145*
Fischer, A. (10) 161, *172*
Fischler, I. (39) *213*
Fiti, M. (14) 2, *26*
Fleischauer, P. D. (2) 60, *71;* (28) 84, *90;* (4) 189, *212*
Flood, T. C. (43) 129, *131*
Flynn, C. M. (4) 61, *71;* (5) 61, *71*
Folting, K. (25) 164, *173;* (20) 179, 180, *187*
Foote, R. S. (3) 174, 179, *186;* (16) 163, *173*
Ford, K. H. (24) 74, 77, *90*
Ford, P. C. 73; (1) 73, *89;* (7) 73, 77, *89;* (8) 73, *89;* (9) 73, 74, 77, 82, *89;* (10) 73, 74, 81, 82, 88, *89;* (11) 73, 86, 88, *89;* (12) 73, 75, *89;* (13) 73, 74, 77, 79, *89;* (14) 73, 77, 87, *89;* (15) 73, 75, 82, 88, *89;* (19) 74, 75, 85, 87, *89;* (24) 74, 77, *90;* (29) 84, *90;* (33) 74, *90;* (6) *212;* (13) 190–192, *212*
Ford-Smith, M. H. (49) 20, 27
Forster, L. S. (30) 9, *26;* (11) 93, *113*
Fox, M. (26) 78, 81, *90*
Frankel, E. N. (23) 191, *212;* (24) 191, *212*

Frankel, L. S. (28) 107, *114*
Frazier, C. (34) 139, *145*
Frazier, C. C. (37) 139, 141, *145*
Freeland, B. H. (30) 135, *145*
Friedman, L. (38) 183, *187;* (39) 183, *187*
Fukuzumi, K. (18) 175, 178, *187*

G

Gafney, H. D. (1) 1, *25;* (6) 73, 77, *89*
Gaibel, Z. L. F. (24) 179, 182, *187*
Gambino, O. (77) 192, *214;* (78) 192, *214*
Gandolfi, M. T. (21) 104, *114*
Garnier, F. (44) 191, *213*
Gaunder, R. G. (64) 13, 27
Geoffroy, G. L. 132; (4) 46, *56;* (16) 190, *212;* (17) 190, *212;* (18) 190, *212;* (38) 141, *145;* (43) 143, 144, *146*
George, T. A. 147; (1) 147, 152, *156;* (3) 147, 154, *156;* (4) 148, 149, 154, *156;* (5) 147, 154, *156;* (8) 148, *157;* (11) 148, *157;* (13) 148, *157;* (17) 149, *157;* (20) 149, *157;* (25) 154, 156, *157*
Ghielmi, S. (1) 46, *56*
Giannotti, C. (24) 116, 121, *131;* (12) 190–192, *212;* (47) 191, *213*
Gilbert, J. C. (45) 183, *187*
Ginley, D. S. (10) 116, 121, 123, *130;* (13) 116, 121, 123, *130;* (37) 139, 141, *145;* (9) 190–192, 201, 210, *212;* (22) 190, *212;* (81) 201, 211, *214*
Giordano, P. J. (15) 116, 130; **189;** (8) 190, *212*
Gladfelter, W. L. (43) 143, 144, *146*
Goe, G. L. (23) 164, *173;* (11) 174, 179, *187*
Goldschmidt, Z. (44) 183, *187*
Gordon, B. M. (48) 20, 27
Gordon, J. G., II (7) 52, *56;* (8) 52, *56;* (9) 52, *56*
Gorman, A. A. (13) 161, *173;* (14) 161, *173*
Gorton, E. M. (16) 94, *114*
Gould, E. S. (61) 13, 27
Graham, M. A. (5) 115, *130*
Grätzel, M. (43) 17, *26*
Gray, H. B. (4) 45, *56;* (5) 46, 47, 49, 51, *56;* (7) 52, *56;* (8) 52, *56;* (9) 52, *56;* (10) 53, *56;* (11) 53, *56;* (13) 53, *56;* (9) 116, *130;* (16) 116, *131;* (15) 148, *157;* (16) 190, *212;* (31) 191, *213;* (32) 191, *213;* (35) 139, *145;* (37) 139, 141, *145*
Gray, H. G. (21) 151, *157*
Greeley, R. H. (9) 174, 179, *187*
Green, M. (9) 132, 133, *145*
Green, M. L. H. (47) 191, *213*
Grevels, F. W. (2) 115, *130;* (5) 189, *212*

Grutsch, P. A.   (27) 164, *173;* (34) 166, 168, *173;* (36) 166, *173;* (45) 170, *173*
Guggenberger, L. J.   (23) *157*
Gutierrez, A. R.   (20) 104, *114*
Gutmann, V.   (32) 111, *114*
Guttel, G.   (25) 77, 78, *90*
Guy, R. G.   (26) 135, *145*

H

Hager, G. D.   (31) 9, *26;* (16) 68, 72
Haines, R. J.   (21) 116, *131*
Haller, I.   (7) 174, 179, *186*
Halpern, J.   (54) 12, 22, *27;* (20) 190, *212*
Hammond, G. S.   (7) 1, *26;* (4) 46, *56;* (5) 46, 47, 49, 51, *56;* (11) 53, *56;* (13) 53, *56;* (9) 116, *130;* (21) 151, *157;* (10) 161, *172;* (11) 161–163, *172;* (12) 161, *173;* (16) 190, *212;* (31) 191, *213;* (32) 191, *213*
Hanes, R. M.   (49) 172, *173*
Harbourne, D. A.   (14) 132, 134, *145;* (15) 132, 134, *145;* (16) 132, 134, *145;* (55) 192, 201, *213;* (56) 192, 201, *213;* (58) 192, 201, *213*
Harrigan, R. W.   (24) 3, *26;* (16) 68, 72
Harrington, J. S.   **57;** (6) 61, 67, *72;* (3) 73, *89;* (42) 169–171, *173*
Harris, E. W.   (8) 1, 8, *26*
Harrod, J. F.   (95) 207, *214*
Hart, E. J.   (27) 81, *90*
Hasso, S.   (70) 192, *214*
Hatchard, C. G.   (34) 117, *131;* (32) 136, *145;* (96) 211, *214*
Hattaway, J. H., Jr.   (24) 179, 182, *187*
Hausser, J. W.   (28) 182, *187;* (29) 182, *187*
Hautala, R. R.   (48) 172, *173*
Hawkins, C. J.   (18) 96, *114*
Head, R. A.   (16) 148, 149, *157*
Heath, G. A.   (2) 147, *156;* (18) 149, *157*
Heck, R. F.   (46) 143, 144, *146;* (47) 143, 144, *146;* (48) 143, *146*
Heidt, L. J.   (3) 159, *172*
Hellier, M.   (80) 195, *214*
Hemingway, R. E.   (15) 2, 5, 7, 23, *26*
Henderson, R.   (43) 143, 144, *146*
Henry, J. P.   (16) 163, *173;* (2) 174, 179, *186;* (3) 174, 179, *186*
Henry, M. S.   **91;** (3) 91–95, 100, 103, 104, 108, 109, *113;* (5) 92, *113;* (6) 92, 94, 102, 103, 105, *113;* (25) 105, 107, 111, *114*
Hercules, D. M.   (19) 2, *26*
Hermann, H.   (7) 115, *130*
Herzog, S.   (18) 92, 95, *113*
Hickey, J. P.   (39) 127, *131*
Hidai, M.   (6) 148, *157;* (7) 148, *157;* (10) 148, *157*
Hill, B.   (42) 191, *213;* (43) 191, *213*

Hill, K. L.   (47) 183, *187*
Hintze, R. E.   (1) 73, *89;* (9) 73, 74, 77, 82, *89;* (12) 73, 75, *89;* (13) 73, 74, 77, 79, *89*
Hipps, K. W.   (25) 3, *26*
Hirashima, T.   (50) 185, *188*
Hoffman, M. Z.   **91;** (3) 91–95, 100, 103, 104, 108, 109, *113;* (5) 92, *113;* (17) 95, *114;* (25) 105, 107, 111, *114;* (41) 169, 171, *173*
Hoffmann, R.   (30) 108, 111, *114;* (22) 152, 154, 156, *157*
Hoggard, P. H.   (10) 93, 104, 108, *113*
Hooper, N. E.   (18) 149, *157*
Hoselton, M.A.   (40) 7, 12–15, 21, *26*
Houk, L. W.   (38) 127, *131*
Hudson, A.   (14) 190–192, *212*
Hudson, B.   (65) 192, *213*
Hudson, R. A.   (49) 185, *187*
Hughey, J. L., IV   (11) 116, 121, *130;* (12) 116, 121, *130;* (11) 190–192, *212*
Huie, B. T.   (24) 135, *145*
Hunter, D. L.   (8) 132, 133, *145*
Hunter, H. R., Jr.   (16) 94, *114*
Hwang, D. R.   (49) 185, *187*

I

Ibers, J. A.   (15) 174, *187;* (16) 174, *187*
Isci, H.   (12) 53, *56*
Ishaq, M.   (20) 116, *131*
Ishino, Y.   (50) 185, *188*
Iske, S. D. A., Jr.   **147;** (1) 147, 152, *156;* (3) 147, 154, *156;* (4) 148, 149, 154, *156;* (14) 148, 149, *157*

J

Jamieson, M. A.   (13) 93, *113;* (25) 105, 107, 111, *114*
Janz, G. J.   (27) 106, 109, *114*
Jardine, F. H.   (19) 163, *173;* (38) 169, *173*
Jenkins, S. H.   (8) 65, 72
Jennings, W.   (42) 191, *213;* (43) 191, *213*
Jesson, J. P.   (23) *157*
Johnson, B. F. G.   (5) 132, 133, *145;* (6) 132, 133, *145;* (17) 132, 134, *145;* (42) 143, 144, *146;* (49) 192, 201–203, *213;* (50) 192, 201, *213;* (59) 192, 201, 211, *213;* (67) 192, *213;* (68) 192, *213;* (93) 203, 205, *214;* (100) 203, *214*
Johnston, R. D.   (100) 203, *214*
Josty, P. L.   (59) 192, 201, 211, *213;* (100) 203, *214*

K

Kaesz, H. D.   (24) 135, *145;* (44) 143, 144, *146;* (94) 203, 205, *214*
Kane, V. V.   (62) 192, *213*

Kane-Maquire, N. A. P. (1) 91, 94, *113;* (2) 91, 94, 96, 107, *113;* (18) 96, *114*
Kanetrowitz, E. R. (17) 95, *114*
Kaplan, B. E. (5) 174, 179, *186*
Kapoor, P. N. (18) 116, *131;* (19) 116, *131*
Kapoor, R. N. (18) 116, *131;* (19) 116, *131*
Kasha, M. (18) 69, *72*
Kawanisi, M. (10) 174, 179, *187*
Kearns, D. R. (23) 105, *114*
Keene, F. R. (34) 11, *26*
Keister, J. B. (71) 192, 210, *214;* (72) 192, 210, *214*
Kelland, J. W. (5) 132, 133, *145;* (17) 132, 134, *145*
Kelly, J. M. (7) 115, *130*
Kelm, H. (22) 105, *114*
Kemp, T. J. (8) 65, *72;* (23) 116, *131*
Kende, A. S. (44) 183, *187*
Kilty, P. A. (93) 203, 205, *214*
King, R. B. (18) 116, *131;* (19) 116, *131;* (20) 116, *131;* (28) 117, *131;* (38) 127, *131;* (1) 132, *145;* (49) 172, *173*
Kirk, A. D. (10) 93, 104, 108, *113*
Kirmse, W. (42) 183, *187;* (43) 183, *187*
Kirsch, P. (33) 191, *213*
Kirsch, P. P. (13) 1, 14, 25, *26*
Klassen, D. M. (14) 67, *72;* (15) 67, *72*
Kliger, D. S. (7) 1, *26;* (13) 53, *56*
Knight, J. (97) 203, *214*
Knobler, C. B. (24) 135, *145*
Knox, S. A. R. (10) 132–134, *145;* (11) 132–134, *145;* (12) 132, 133, *145;* (13) 132, 133, *145;* (51) 192, 201, *213;* (52) 192, 201, *213;* (53) 192, 201, *213;* (54) 192, 201, *213;* (94) 203, 205, *214*
Kochi, J. K. (13) 174, *187;* (19) 179, 181, *187;* (20) 179, 180, *187;* (21) 164, *173;* (24) 164, *173;* (25) 164, *173*
Kodama, T. (6) 148, *157*
Koepke, J. W. (94) 203, 205, *214*
Koerner von Gustorf, E. (5) 189, *212;* (39) *213*
König, E. (8) 92, 95, *113*
Koser, G. F. (51) 186, *188*
Kosturik, J. M. (40) 183, *187*
Kotz, J. C. (39) 142, *146*
Krausz, P. (44) 191, *213*
Krentzien, H. (47) 20, *26*
Kropp, P. J. (24) 179, 182, *187*
Kruczynski, L. (7) 132, 133, *145*
Kursten, G. (29) 8, *26*
Kutal, C. 158; (27) 164, *173;* (28) 162, 164, *173;* (29) 164–166, *173;* (33) 166, *173;* (34) 166, 168, *173;* (36) 166, *173;* (45) 170, *173;* (47) 171, *173;* (52) 179, *188;* (53) 179, *188*

L

Laine, R. M. (13) 190–192, *212*
Lakshminarayanan, G. R. (27) 106, 109, *114*
Lamola, A. A. (46) 170, *173*
Lane, J. E. (2) 158, *172*
Langford, C. H. (1) 91, 94, *113;* (2) 91, 94, 96, 107, *113;* (28) 107, *114;* (31) 109, *114*
Lappert, M. F. (14) 190–192, *212*
Latimer, W. M. (38) 12, *26*
Laurence, G. A. (4) 1, *25*
Laurence, G. S. (5) 1, *25*
Lee, C. S. (16) 94, *114*
Lee, J. T. 28; (2) 73, *89*
Leigh, G. J. (2) 147, *156;* (16) 148, 149, *157;* (18) 149, *157*
L'Eplattenier, F. (92) 203, 204, *214;* (99) 203, *214*
LePort, P. C. (64) 192, *213*
Lever, A. B. P. (32) 74, *90*
Lewis, J. (5) 132, 133, *145;* (6) 132, 133, *145;* (17) 132, 134, *145;* (42) 143, 144, *146;* (49) 192, 201–203, *213;* (50) 192, 201, *213;* (59) 192, 201, 211, *213;* (67) 192, *213;* (68) 192, *213;* (93) 203, 205, *214;* (100) 203, *214*
Lewis, N.S. (6) 51, 53, *56;* (8) 53, *56;* (9) 52, *56;* (10) 53, *56;* (11) 53, *56;* (13) 53, *56*
Leyland, R. L. (13) 161, *173;* (14) 161, *173*
Libienthal, J. (37) 139, 141, *145*
Lichtenberger, D. L. (25) 117, 123, 126, *131*
Liengme, B. V. (14) 132, 134, *145;* (16) 132, 134, *145;* (56) 192, 201, *213;* (58) 192, 201, *213*
Lin, C.-T. (11) 1, 5, 15, *26;* (32) 7, 10, 15, 20, 22, *26;* (40) 1, 12–15, 21, *26*
Lip, H. (18) 96, *114*
Little, J. (48) 172, *173*
Liu, R. S. H. (32) 117, *131*
Livingston, R. S. (3) 159, *172*
Longoni, G. (3) 132, *145*
Lytle, F. E. (19) 2, *26*

M

Maestri, M. (43) 17, *26;* (10) 66, *72;* (3) 91–95, 100, 103, 104, 108, 109, *113;* (4) 91–93, 103, *113;* (5) 92, *113;* (9) 93, *113;* (12) 93, 111, *113;* (13) 93, *113*
Mahon, C. (23) 3, 4, *26*
Mehta, B. D. (15) 94, *113*
Majerski, Z. (36) 182, 183, *187*
Malatesta, L. (1) 46, *56*
Malin, J. M. (30) 87, *90*
Malouf, G. (10) 73, 74, 81, 82, 88, *89;* (11) 73, 86, 88, *89;* (15) 73, 75, 82, 88, *89;* (16) 73, *89;* (29) 84, *90;* (6) 190, *212*

Manabe, O. (50) 185, *188*
Manfrin, M. F. (5) 1, *25*
Mann, C. D. M. (30) 135, *145*
Mann, F. G. (37) 169, *173*
Mann, K. R. (3) 46, 51, 56; (4) 46, 56; (5) 46, 47, 49, 51, 56; (7) 52, 56; (8) 52, 56; (9) 52, 56; (10) 53, 56; (11) 53, 56; (13) 53, 56
Manning, A. R. (31) 135, *145*
Manuel, T. A. (61) 192, *213*
Marchand, A. P. (323) 182, 183, *187*
Marcus, R. A. (35) 11, 26; (36) 11, 26
Markham, J. (44) 170, *173*
Marko, L. (19) 134, 135, 141, *145;* (33) 138, *145*
Marshall, J. A. (23) 179, 182, *187*
Martin, J. L. (7) 132, 133, *145*
Martinengo, S. (29) 135, *145*
Mason, W. R. (12) 53, *56*
Math, K. (43) 191, *213*
Matheson, M. S. (17) 69, 72
Matheson, T. W. (23) 135, *145*
Matsubara, T. (18) 73–75, 78–80, *89;* (20) 76, 80, *89*
Mau, A. W.-H. (2) 158, *172*
Mayer, G. E. (27) 106, 109, *114*
Mays, M. J. (28) 135, *145;* (97) 203, *214*
McBride, E. P. (8) 1, 8, *26*
McCleverty, J. A. (74) 192, *214;* (75) 192, *214*
McDonald, D. P. (7) 73, 77, *89;* (9) 73, 74, 77, 82, *89*
McKeon, J. E. (2) 174, 179, *186;* (3) 174, 179, *186;* (16) 163, *173*
McMillian, W. G. (6) 161, *172*
McNelis, E. (45) 191, *213;* (46) 191, *213*
Meakin, P. (23) *157*
Meinwald, J. (5) 174, 179, *186*
Meisel, D. (17) 69, 72
Memming, R. (28) 8, *26;* (29) 8, *26*
Merkel, P. B. (23) 105, *114*
Merle, G. (12) 190–192, *212;* (24) 116, 121, *131*
Meyer, T. J. (2) 1, *25;* (10) 1, *26;* (18) 2, 17, *26;* (26) 3, 5, 7, *26;* (34) 11, *26;* (46) 20, *26; 28;* (12) 67, *72;* (2) 73, *89;* (11) 116, 121, *130;* (12) 116, 121, *130;* (36) 121, *131;* (11) 190–192, *212*
Milone, L. (66) 192, *213*
Miller, M. A. (7) 161, *172*
Miskowski, V. M. (11) 53, *56;* (13) 53, *56*
Misona, A. (7) 148, *157;* (27) 191, *212*
Moaz, N. (25) 191, *212;* (26) 191, *212*
Moggi, L. (5) 1, *25;* (10) 66, 72; (3) 91–95, 100, 103, 104, 108, 109, *113;* (5) 92, *113;* (12) 93, 111, *113;* (21) 104, *114;* (41) 169, 171, *173*
Monroe, B. M. (32) 117, *131*
Morse, D. L. (22) 76, 77, *90;* (7) 190, *212;* (16) 116, *131;* (22) 190, *212*

Morse, K. W. (35) 166, *173*
Moses, F. G. (32) 117, *131*
Muetterties, E. L. (30) 108, 111, *114;* (23) *157*
Mulac, W. A. (17) 69, 72
Murdoch, H. D. (98) 203, *214*
Murov, S. (11) 161–163, *172*
Murphy, M. A. (41) 127, 128, *131*

N

Nasielski, J. (33) 191, *213*
Navon, G. (3) 1, 5, 15, *25*
Nelson, H .H., III (41) 127, 128, *131*
Neumann, H. M. (16) 94, *114;* (26) 106, 107, *114*
Nicholson, B. K. (14) 190–192, *212*
Nikoletić, M. (36) 182, 183, *187*
Nishiguchi, T. (18) 175, 178, *187*
Nisikawa, Y. (10) 174, 179, *187*
Nobinger, G. L. (13) 53, *56*
Noda, K. (50) 185, *188*
Norbury, A. H. (37) 125, *131*
Nordmeyer, F. (60) 13, *27;* (34) 74, *90*
Noyori, R. (10) 174, 179, *187*
Nozaki, H. (10) 174, 179, *187*
Nyholm, R. S. (21) 116, *131;* (36) 139, *145;* (90) 203, *214;* (91) 203, *214;*

O

O'Brien, R. J. (30) 135, *145*
Ogata, I. (27) 191, *212*
Ogura, T. (50) 185, *188*
Oliver, B. G. (27) 106, 109, *114*
Olson, K. R. (41) 169, 171, *173*
Orhanovic, M. (45) 20, *26*
Orio, A. A. (4) 46, *56*
Orlowski, R. C. (40) 183, *187*
Ostazewski, A. P. P. (30) 165, *173;* (31) 165, *173*
Ostella, D. (66) 192, *213*
Otteson, D. K. (16) 116, *131*

P

Paeonessa, R. S. **189**
Palyi, G. (19) 134, 135, 141, *145*
Papaconstantinou, E. (17) 95, *114*
Pappas, P. R. (51) 186, *188*
Parham, W. E. (27) 182, *187*
Parker, C. A. (32) 136, *145;* (34) 117, *131;* (96) 211, *214*
Pdungsap, L. (22) 76, 77, *90*
Peake, B. M. (40) 142, *146*
Pearson, R. G. (11) 66, 72; (35) 121, *131*
Peet, W. G. (23) *157*
Penfold, B. R. (20) 134, 135, 141, *145*
Peng, H. M. (44) 130, *131*
Peraldo, M. (27) 135, *145*
Pereira, M. S. (30) 87, *90*
Perutz, R. N. (5) 115, *130;* (6) 116, *130*

Peters, H. M. (23) 191, *212*
Petersen, J. D. (1) 73, *89*; (9) 73, 74, 77, 82, *89*; (15) 73, 75, 82, 88, *89*; (33) 74, *90*
Petersen, J. V. (39) 142, *146*
Piacenti, F. (19) 134, 135, 141, *145*
Pickett, C. J. (16) 148, 149, *157*
Pierantozzi, P. (18) 190, *212*
Pillsbury, D. (43) 191, *213*
Pitts, J. N. (33) 117, *131*
Platbrood, G. (35) 191, *213*; (36) 191, *213*; (37) 191, *213*
Poë, A. (87) 203, *214*; (88) 203, *214*
Poliakoff, M. (5) 115, *130*; (28) 191, *212*; (29) 191, *212*
Pompe, A. (2) 158, *172*
Pook, K.-H. (42) 183, *187*; (43) 183, *187*
Porter, G. B. (10) 93, 104, 108, *113*; (14) 94, *113*
Przystas, T. J. (51) 13, 20, *27*
Purdie, D. (37) 169, *173*

Q

Quin, H. W. (18) 163, *173*

R

Rabani, J. (17) 69, *72*
Rabinowitch, E. (3) 159, *172*
Reardon, E. J., Jr. (24) 179, 182, *187*
Reed, H. W. B. (26) 135, *145*
Reed, R. C. (39) 142, *146*
Rehani, S. K. (5) 132, 133, *145*; (17) 132, 134, *145*
Reichard, D. (17) 116, 120, *131*
Rest, A. J. (5) 115, *130*
Reynolds, W. L. (23) 3, 4, *26*
Richards, R. L. (2) 147, *156*
Rietvelde, D. (38) *213*
Risby, T. H. (38) 141, *145*
Roberts, J. D. (33) 182, 183, *187*; (34) 182, 183, *187*
Roberts, P. J. (57) 192, 201, *213*
Robinson, B. H. (20) 134, 135, 141, *145*; (23) 135, *145*; (40) 142, *146*
Rockley, M. C. (10) 93, 104, 108, *113*
Rodgers, M. A. J. (14) 161, *173*
Rogers, R. D. (36) 166, *173*
Rosenberg, E. (43) 129, *131*
Rossi, A. R. (30) 108, 111, *114*
Rotella, F. J. (32) 165, *173*
Rule, L. (38) 169, *173*

S

Sacco, A. (1) 46, *56*; (12) 148, *157*
Saillant, R. B. (94) 203, 205, *214*
Saji, T. (16) 2, 7, *26*; (33) 10, *26*
Salah, O. M. A. (32) 165, *173*
Salet, C. (22) 3, *26*
Salmon, D. J. (18) 2, 17, *26*
Salomon, M. F. (26) 164, *173*; (37) 182, *187*

Salomon, R. G. (26) 164, *173*; **174**; (13) 174, *187*; (17) 175, *187*; (19) 179, 181, *187*; (20) 179, 180, *187*; (22) 179, 182, *187*; (37) 182, *187*
Sams, J. R. (14) 132, 134, *145*; (16) 132, 134, *145*; (56) 192, 201, *213*; (58) 192, 201, *213*
Sandler, S. R. (26) 182, *187*
Sandrini, D. (21) 104, *114*
Santambrogio, E. (25) 135, *145*
Saran, M. S. (19) 116, *131*
Sarhangi, A. (43) 129, *131*
Sasse, W. H. F. (2) 158, *172*
Satori, G. (28) 84, *90*
Scandola, F. (52) 20, *27*; (7) 92, *113*; (19) 103, *114*; (41) 169, 171, *173*
Schanck, L. G. (28) 182, *187*; (29) 182, *187*
Schenck, G. O. (25) 179, 182, *187*
Schleyer, P. von R. (35) 182, 183, *187*
Schmid, R. (32) 111, *114*
Schrage, K. (60) 192, *213*
Schroeder, M. A. (22) 190, *212*; (30) 191, 197, 203, 210, 211, 212, *212*; (34) 191, *213*; (40) 191, *213*; (41) 191, 192, 203, 207, 208, 210–212, *213*
Schultenover, D. G. (41) 183, *187*
Schwarz, H. (40) 7, 12–15, 21, *26*
Schweizer, E. E. (27) 182, *187*
Schwendiman, D. P. (27) 164, *173*; (28) 162, 164, *173*; (29) 164–166, *173*; (52) 178, *188*; (53) 179, *188*
Siebold, C. D. (8) 148, *157*; (11) 148, *157*
Serpone, N. (10) 66, *72*; (12) 93, 111, *113*; (13) 93, *113*; (25) 105, 107, 111, *114*
Servis, K. L. (33) 182, 183, *187*
Seyferth, D. (21) 134, 135, 141, *145*
Shapley, J. R. (71) 192, 210, *214*; (72) 192, 210, *214*; (73) 192, *214*; (79) 192, *214*
Shaw, B. L. (26) 135, *145*
Shaw, G. (89) 203, *214*
Shechter, H. (38) 183, *187*; (39) 183, *187*
Sheline, R. K. (4) 115, *130*; (36) 139, *145*
Shiron, M. (25) 77, 78, *90*
Shortland, A. C. (85) 203, 205, *214*
Shubkin, R. L. (17) 116, 120, *131*
Siefert, E. E. (84) 203, *214*
Silber, H. B. (37) 12, *26*
Simpson, J. (40) 142, *146*
Skell, P. S. (26) 182, *187*
Sloan, T. E. (29) 117, 123, 124, *131*
Smith, J. A. (39) 183, *187*
Smith, P. G. (14) 161, *173*
Smith, R. F. (44) 183, *187*
Snyder, R. G. (15) 174, *187*; (16) 174, *187*
Solomon, B. S. (15) 163, *173*

Solomon, R. G. (21) 164, *173;* (24) 164, *173;* (25) 164, *173*
Sprintschnik, G. (13) 1, 14, 25, *26*
Sprintschnik, H. W. 28; (13) 1, 14, 25, *26;* (2) 73, *89*
Spurling, T. H. (2) 158, *172*
Srinvasan, R. (17) 163, *173;* (22) 164, *173;* (4) 174, 175, 179, *186;* (6) 174, 179, *186;* (7) 174, 179, *186;* (21) 179, *187*
Steel, S. (15) 163, *173*
Steinmetz, R. (25) 179, 182, *187*
Sterling, R. (33) 166, *173*
Stiddard, M. H. B. (21) 116, *131*
Stolz, I. W. (4) 115, *130*
Stone, F. G. A. (10) 132–134, *145;* (11) 132–134, *145;* (12) 132, 133, *145;* (13) 132, 133, *145;* (51) 192, 201, *213;* (52) 192, 201, *213;* (53) 192, 201, *213;* (54) 192, 201, *213;* (86) 203, 210, *214;* (89) 203, *214*
Streib, W. E. (25) 164, *173;* (20) 179, 180, *187*
Strouse, C. E. (41) 142, *146*
Stuermer, D. H. (7) 73, 77, *89;* (8) 73, *89;* (9) 73, 74, 77, 82, *89*
Sunko, D. E. (36) 182, 183, *187*
Sutin, N. 1; (3) 1, 5, 15, *25;* (9) 1, 21, *26;* (11) 1, 5, 15, *26;* (12) 1, 6, *26;* (17) 2, 15, 17, *26;* (27) 5, 7, 15, 18, *26;* (32) 7, 10, 15, 20, 22, *26;* (40) 7, 12, 13, 14, 15, 21, *26;* (44) 22, 23, 25, *26;* (45) 20, *26;* (48) 20, *27;* (49) 20, *27;* (51) 13, 20, *27;* (53) 13, 20, 22, *27;* (4) 73, *89;* (23) 77, *90;* (43) 170, *173*
Sutton, P. W. (22) 135, *145*
Sweet, E. M. (47) 171, *173;* (48) 172, *173;* (49) 172, *173*
Swift, E. H. (39) 12, *26*
Sykes, A. G. (59) 13, *27*

T

Tachi, K. (18) 175, 178, *187*
Tachikawa, M. (79) 192, *214*
Takats, J. (7) 132, 133, *145*
Tang Wong, K. L. (48) 191, *213*
Taube, H. (46) 20, *26;* (47) 20, *26;* (58) 13, *27;* (60) 13, *27;* (62) 13, *27;* (64) 13, *27;* (34) 74, *90*
Tham, W. S. (23) 135, *145*
Thich, J. A. (2) 46, *56;* (10) 53, *56*
Thomas, J. L. (48) 191, *213*
Tokel-Takvoryan, N. E. (15) 2, 5, 7, 23, *26*
Todd, L. J. (39) 127, *131*
Toma, H. E. (41) 15, *26*
Tominari, K. (7) 148, *157;* (10) 148, *157*
Trecker, D. J. (16) 163, *173;* (1) 174, 179, *186;* (2) 174, 179, *186;* (3) 174, 179, *186*
Treichel, P. M. (17) 116, 120, *131;* (22) 116, *131*

Trimm, D. L. (54) 13, 22, *27*
Trogler, W. C. (35) 139, *145*
Trotter, J. (57) 192, 201, *213*
Tsai, J. H. (18) 163, *173*
Tunstall, S. M. (8) 65, *72*
Turner, J. J. (5) 115, *130;* (6) 115, *130;* (28) 191, *212*
Turner, R. F. (5) 115, *130*
Turro, N. J. (10) 161, *172;* (12) 161, *173*
Twigg, M. V. (6) 132, 133, *145;* (42) 143, 144, *146;* (49) 192, 201–203, *213;* (50) 192, 201, *213;* (87) 203, *214;* (88) 203, *214*
Tyler, D. R. (37) 139, 141, *145*

U

Uchida, T. (6) 148, *157*
Uchida, Y. (6) 148, *157;* (7) 148, *157;* (10) 148, *157*
Ullah, S. S. (59) 192, 201, 211, *213*
Underhill, M. (69) 192, *214;* (70) 192, *214*

V

Vaglio, G. A. (66) 192, *213;* (76) 192, *214;* (77) 192,*214;* (78) 192, *214;*
Valle, M. (66) 192, *213;* (77) 192, *214*
van Bronswijk, W. (90) 203, *214*
Van-Catledge, F. A. (8) 161, *172*
Van Dine, G. S. (35) 182, 183, *187*
Van Houten, J. (6) 1, 11, *26;* (21) 2, *26;* **57;** (3) 60, 68, *71;* (6) 61, 67, *67;* (7) 65, 66, 68, *72;* (9) 65, 66, *72;* (3) 73, *89;* (42) 169–171, *173*
Van Meter, F. M. (26) 106, 107, *114*
Varani, G. (52) 20, *27;* (7) 92, *113;* (41) 169, 171, *173*
Vaudo, A. F. (17) 95, *114*
Venanzi, L. M. (14) 174, *187*
Voecks, G. (43) 191, *213*
Vogel, M. (34) 182, 183, *187*
Vohra, A. G. (38) 169, *173*
von Gustorf, E. K. (2) 115, *130;* (7) 115, *130*

W

Wagner, S. D. (4) 148, 149, 154, *156;* (19) 149, *157*
Walters, R. T. (20) 104, *114*
Warrener, R. N. (8) 174, 179, *187*
Wasgestian, F. (22) 105, *114*
Watson, D. J. (40) 142, *146*
Watts, R. J. (6) 1, 11, *26;* (21) 2, *26;* (31) 9, *26;* **57;** (3) 60, 68. *71;* (6) 61, 67, *72;* (7) 65, 66, 68, *72;* (9) 65, 66, *72;* (3) 73, *89;* (29) 84, *90;* (31) *90;* (33) 74, *90;* (39) 171, *173;* (42) 169–171, *173*
Weaver, T. R. (13) 67, *72*
Webster, D. E. (65) 192, *213*

Weiss, E.   (98) 203, *214*
Weller, A.   (15) 163, *173*
Wells, A. F.   (37) 169, *173*
Wells, P. B.   (65) 192, *213*
Westlake, D. J.   (9) 132, 133, *145*
Whipple, E. B.   (1) 174, 179, *186*
Whitesides, G. M.   (23) 164, *173;* (11) 174, 179, *187*
Whiting, M. C.   (62) 192, *213*
Whitten, D. G.   (2) 1, *25;* (10) 1, *26;* (13) 1, 14, 25, *26;* (26) 3, 5, 7, *26;* **28;** (2) 73, 89; (21) 76, 77, *90*
Wiberg, K. B.   (4) 160, *172*
Wiedemann, W.   (28) 182, *187*
Wilcox, C. F.   (6) 161, *172*
Wilkinson, G.   (20) 164, *173;* (21) 190, *212*
Wilkinson, J. R.   (39) 127, *131*
Willard, K. F.   (24) 179, 182, *187*
Williams, I. G.   (100) 203, *214*
Williams, L. L.   (48) 20, 27
Williams, R. M.   (8) 52, *56*
Wilputte-Steinert, L.   (33) 191, *213;* (35) 191, *213;* (36) 191, *213;* (37) 191, *213;* (38) *213*
Wilt, J. W.   (40) 183, *187;* (41) 183, *187*
Windsor, M. W.   (10) 93, 104, 108, *113*
Winstein, S.   (6) 161, *172*
Winterle, J. S.   (7) 1, *26*
Wojcicki, A.   (29) 117, 123, 124, *131*
Wright, R. E.   (20) 104, *114*

Wrighton, M.   (8) 116, *130;* (9) 116, *130;* (21) 151, *157;* (1) 189, 191, *212;* (22) 190, *212;* (31) 191, *213;* (32) 191, *213;* (34) 191, *213*
Wrighton, M. S.   (22) 76, 77, *90;* (1) 115, 116, 130, *130;* (10) 116, 121, 123, *130;* (13) 116, 121, 123, *130;* (15) 116, *130;* (16) 116, *131;* (4) 132, *145;* (18) 134, *145;* (37) 139, 141, *145;* (45) 143, 144, *146;* (44) 170, *173;* **189;** (2) 189, *212;* (7) 190, *212;* (8) 190, *212;* (9) 190–192, 201, 210, *212;* (15) 190–192, *212;* (30) 191, 197, 203, 210–212, *212;* (40) 191, *213;* (41) 191, 192, 203, 207, 208, 210–212, *213;* (81) 201, 211, *214*
Wyatt, P.   (12) 161, *173*

Y

Young, R. C.   (10) 1, *26;* (34) 11, *26*
Young, R. J.   (18) 2, 17, *26*
Yu, S.-M.   (51) 186, *188*

Z

Zanella, A.   (24) 74, 77, *90*
Zarnegar, P. P.   (21) 76, 77, *90*
Zawadzki, J. F.   (41) 183, *187*
Zinato, E.   (29) 108, *114*
Zipperer, W. C.   (20) 116, *131*
Zwickel, A.   (58) 13, *27*

SUBJECT INDEX

A

Absorbance against irradiation
   time, plots of 204–206
Absorption
   bands for CpFe(CO)(L)X
     complexes, electronic 118
   and emission spectral data for
     ML_6 complexes 48
   measurements, luminescence and   60
   spectral data, electronic 137, 153
   spectroscopy, ground state 95
   spectrum (a) 94
     of $Cr(bpy)_3^{3+}$ 96, 97
     electronic 136, 141
       of $CH_3CCo_3(CO)_9$ 136
       of $CpFe(CO)_2Br$ 119
       of $CpFe(CO)_2Cl$ 119
       of $CpFe(CO)_2I$ 119
       of $Cu[P(C_6H_5)_3]_2BH_4$ and
         $P(C_6H_5)_3$ 168
       of dinitrogen, alkyldiazenido,
         and alkylhydrazido
         derivatives of
         molybdenum 150
       of $MoBr(N_2C_4H_9)(dppe)_2$ 153
       of $MoI(N_2C_8H_{17})(dppe)_2$-
         BF_4 155
       of $Mo(N_2)_2(dppe)_2$ 150
       of $M_3(CO)_{12}$ 192, 193
       of $Ru(bpy)_3^{2+}$ 2
Acceptors, quenching by neutral .. 35
Acceptors, quenching by positively
   charged 36
Alcohol, allyl 185
Alkene(s)
   isomerization ....:.. 192
     $C_3(CO)_{12}$ photocatalyzed ... 194
   selectivity for terminal 197
   –silane ratio, product distribu-
     tion variation with 197
   with silanes, reaction of 195
   with silanes using trinuclear
     metal carbonyl catalyst
     precursors, photocatalyzed
     reactions of 189
Alkyldiazenido complexes, photochemical aspects of the preparation of 149
Alkylhydrazido derivatives of
   molybdenum, electronic
   absorption spectra of dinitrogen, alkyldiazenido, and . 150
Alkylidyne clusters, iron 134

Allenes 182
Allyl alcohol 185
Allylic rearrangement 175
Allylrhodium hydride 177
Amine complexes, roles of charge
   transfer states in the photo-
   chemistry of $Ru(II)$ 73
Amines, quenching of $Ru(II)$ com-
   plex excited states by 38
Anions, effect of 105
Association mechanism 93

B

2,2′-Bipyridine complexes of
   $Ir(III)$ and $Ru(II)$ 57
Bis(dinitrogen)bis[1,2-bis(diphenylphosphino)ethane]-
   molybdenum, photochemistry
   of 147
Bonds, metal–metal 191

C

$C_3(CO)_{12}$ photocatalyzed alkene
   isomerization 194
η^5-$C_5H_5Mo(CO)_3X$ 115
cis-η^5-$C_5H_5Mo(CO)_2(PPh_3)X$ ... 115
$trans$-η^5-$C_5H_5Mo(CO)_2(PPh_3)X$. 115
$CH_3CCo_3(CO)_9$
   electronic absorption spectra of . 136
   photoinduced declusterification
     of 132
   photolysis of 140
$CH_3CCo_3(CO)_7(PPh_3)_2$, thermal
   reactivity of $CH_3CCo_3(CO)_8$-
   PPh_3 and 141
$CH_3CCo_3(CO)_8PPh_3$ and
   $CH_3CCo_3(CO)_7(PPh_3)_2$,
   thermal reactivity of 141
$CHCl_3$ solution of $Cr(CNIph)_6$,
   degassed 50
$CHCl_3$ solution of $W(CNPh)_6$,
   degassed 50
ClO_4^- and Cl^- aqueous solutions,
   photophysical and photochemical constants for ^2E
   states in concentrated 106
$CpFe(CO)_2Br$, electronic absorption spectra of 119
$CpFe(CO)_2Cl$ 123
   electronic absorption spectra of 119

CpFe(CO)$_2$I, electronic absorption spectra of	119
CpFe(CO)(L)X complexes, electronic absorption bands for	118
CpFe(CO)$_2$NCS, IR spectra of	124
CpFe(CO)$_2$SCN, IR spectra of	124
CpFe(^{12}CO)$_2$X and CpFe(^{12}CO)-(^{13}CO)X, carbonyl IR spectral data for	119
CpFe(CO)$_2$X + PPh$_3$ → CpFe(CO)(PPh$_3$)X + CO, quanum yields for the reactions	120
CpMo(CO)$_2$(PPh$_3$)X, quantum yields for reactions of	126
CpMo(CO)$_n$(PPh$_3$)$_{3-n}$X complexes, spectral data for	125
CpMo(CO)$_3$X + PPh$_3$ → cis-CpMo(CO)$_2$(PPh$_3$)X, quantum yields for the reaction	126
Cr(III), solution medium effects on the photophysics and photochemistry of polypyridyl complexes of	91
Cr$_{aq}^{2+}$ and Eu$_{aq}^{2+}$, rate constants for outersphere reductions by	13
Cr(bpy)$_3^{3+}$	91
absorption spectrum of	96, 97
energy level diagram of	92
excitation spectrum (uncorrected) of the phosphorescence of	98
phosphorescence emission intensity of	99
phosphorescence emission spectrum of	98
quantum yields for photoreaction of	103
Cr(CNIph)$_6$, degassed CHCl$_3$ solution of	50
Cr(CNIph)$_6$, degassed pyridine solution of	47
Cr(CNPh)$_6$	46
Cr(phen)$_3^{3+}$	91
Cu$_{aq}^+$, oxidation of	21
CuCl-sensitized production of Q, quantum yields for	165
Cu[P(C$_6$H$_5$)$_3$]$_2$BH$_4$ and P(C$_6$H$_5$)$_3$, electronic absorption spectra of	168
Carbenium ion(s)	183
copper–	181
cyclopropyl	182
Carbonyl complexes, photochemical processes in cyclopentadienylmetal	115
Catalysis	190
of olefin photoreaction by transition metal salts	174
Catalyst development	160
Catalyst precursors, photocatalyzed reactions of alkenes with silanes using trinuclear metal carbonyl	189
Charge transfer states in the photochemistry of Ru(II) amine complexes, roles of	73
Chemicals	94
Clusters, iron alkylidyne	134
Clusters, organometallic	132
Continuous photolysis	95, 102
Copper	174
–carbenium ion	181
-catalyzed photochemical reactions	178
(I) compounds, sensitization by	163
Cyclobutenes	182
1,5-Cyclooctadiene, photolysis of	176
Cyclopentadienyliron halides	116
Cyclopentadienylmetal carbonyl complexes, photochemical processes in	115
Cyclopropyl carbenium ion	182
Cyclopropylcarbinyl	182

D

Deactivation of the luminescent excited state of Ru(bpy)$_3^{2+}$, free energy diagram for the	9
Declusterification	133
photoinduced	
of CH$_3$CCo$_3$(CO)$_9$	132
of HCCo$_3$(CO)$_9$	132
of HFeCo$_3$(CO)$_{12}$	132
Dihydride, rhodium	178
Dimers, norbornene	181
Dinitrogen, alkyldiazenido, and alkylhydrazido derivatives of molybdenum, electronic absorption spectra of	150

E

^{2}E states in concentrated ClO$_4^-$ and Cl$^-$ aqueous solutions, photophysical and photochemical constants for	126
^{2}E states in water, photophysical and photochemical constants for	104
Eu$_{aq}^{2+}$, rate constants for outersphere reductions by Cr$_{aq}^{2+}$ and	13
Electrochemical measurements	31
Electron	
acceptors, quenching of hydrophobic ruthenium complex luminescence	35–37
donors, quenching and reaction with neutral	38
transfer photochromism	86

SUBJECT INDEX

Electron (*continued*)
  transfer reactions of hydrophobic
    analogs of $Ru(bipy)_3^{2+}$,
    light-induced 28
Emission
  intensity and spectrum of
    $Cr(bpy)_3^{3+}$, phosphores-
    cence 98, 99
  spectra of $Ru(bpy)_3^{2+}$ 69
  spectral data for ML_6 com-
    plexes, absorption and 48
Energy
  conversion, solar 44
  diagram for the deactivation of
    the luminescent excited
    state of $Ru(bpy)_3^{2+}$, free .. 9
  for hydrophobic $Ru(II)$ com-
    plexes, luminescence
    maxima, lifetimes, and
    excited state 33
  level diagram of $Cr(bpy)_3^{3+}$ ... 92
  levels in TiO_2 and polypyridine-
    ruthenium(II) complexes . 6
  solar 158
  storage cycle, photochemical . 158, 159
  storage reaction, use of transition
    metal compounds to
    sensitize an 158
  transfer in double minima
    potential wells 58
  transfer in $Ru(bpy)_3^{2+}$ 66
Excited state(s)
  energies for hydrophobic $Ru(II)$
    complexes, luminescence
    maxima, lifetimes, and 33
  lifetimes 3
  of polypyridine complexes of
    $Ru(II)$ and $Os(II)$,
    properties and reactivities
    of the luminescent 1
  reactivity 11
  reduction potentials 7
  of $Ru(bpy)_3^{2+}$, luminescent 5–9
  spectra 3, 74
  of tris(2,2'-bipyridine)
    ruthenium(II) and
    -osmium(II), properties
    of the 2
  types of 74

F

Fe_{aq}^{2+}, oxidation of 20–22
$Fe_3(CO)_{12}$, comparison of
  $Fe(CO)_5$ and 196
$Fe(CO)_5$- and $Fe_3(CO)_{12}$-photo-
  catalyzed reaction of 1-pentene
  with $HSiEt_3$ 196
$Fe_3(CO)_{12}$ photocatalyzed pentene
  isomerization 195
Flash photolysis 95, 100
Fragmentation 182

G

Gas chromatographic traces 199

H

$HCCo_3(CO)_9$
  irradiation of 138
  photoinduced declusterification
    of 132
  photolysis of 137
HCl, photolysis of $Ru(bpy)_3^{2+}$
  in $1M$ 63
$HFeCo_3(CO)_{12}$ and $HFeCo_3$-
  $(CO)_{10}(PPh_3)_2$, photolysis of 140
$HFeCo_3(CO)_{12}$, photoinduced
  declusterification of 132
$HSiEt_3$
  and *cis*-2-pentene, $M_3(CO)_{12}$-
    photocatalyzed reaction of . 198
  comparison of $Fe(CO)_5$- and
    $Fe_3(CO)_{12}$-photocatalyzed
    reaction of 1-pentene with . 196
  $M_3(CO)_{12}$-photocatalyzed
    reaction of 1-pentene and .. 196
HX solutions, quantum yields for
  the photooxidation of
  $Rh_2(bridge)_4^{2+}$ in 54
Halides, cyclopentadienyliron 116
Halides, molybdenum-carbonyl ... 116
Heat, low-grade ($\sim 100°C$) 160
Hydride, allylrhodium 177
Hydrogen transfer 178
Hydrogenation, transfer 175
Hydrophobic analogs of
  $Ru(bipy)_3^{2+}$, light-induced
  electron transfer reactions of . 28
Hydrosilation, $M_3(CO)_{12}$-photo-
  catalyzed isomerization vs. .. 200
Hydrosilation, relative rates of
  isomerization and 199

I

$Ir(III)$ and $Ru(II)$, 2,2'-bipyridine
  complexes of 57
$Ir(bpy)_3^{2+}$ in $0.01M$ NaOH,
  photolysis of 65
$Ir(bpy)_3^{3+}$ solutions, photolysis
  of $Ru(bpy)_3^{2+}$ and 60
Inorganic sensitizers 164
Intramolecular photocycloaddition . 179
Ions
  carbenium 183
  copper–carbenium 181
  cyclopropyl carbenium 182
IR spectra
  of $CpFe(CO)_2NCS$ 124
  of $CpFe(CO)_2SCN$ 124
  data for relevant complexes .. 202, 203
IR spectral data for $CpFe(^{12}CO)_2X$
  and $CpFe(^{12}CO)(^{13}CO)X$,
  carbonyl 119

Iron alkylidyne clusters 134
Iron complexes 117
Irradiation of $HCCo_3(CO)_9$ 138
Irradiation procedures, general ... 135
Isocyanide complexes, metal– 44
Isomeric mixture 124
Isomerization
  of alkenes 192
  $C_3(CO)_{12}$-photocatalyzed alkene 194
  cis–trans 179
  $Fe_3(CO)_{12}$-photocatalyzed
    pentene 195
  geometric 128
  vs. hydrosilation, $M_3(CO)_{12}$-
    photocatalyzed 200
  and hydrosilation, relative
    rates of 199
  olefin 139
Isomers, linkage 123

L

Lifetimes
  excited state 3
  and excited-state energies for
    hydrophobic Ru(II) com-
    plexes, luminescence maxima 33
  luminescence 66
Ligands with delocalized π
  systems 169–171
Light-induced electron transfer
  reactions of hydrophobic
  analogs of $Ru(bipy)_3^{2+}$ 28
Linkage isomers 123
Low-grade ($\sim 100°C$) heat 160
Luminescence 95
  and absorption measurements .. 60
  by neutral electron acceptors,
    quenching of hydrophobic
    ruthenium complex 35
  by positively charged electron
    acceptors, quenching of
    hydrophobic ruthenium
    complex 37
  spectroscopy 96

M

$M_3(CO)_{12}$ (M=Fe, Ru, Os) 189
  absorption spectra of 192
  disappearance quantum yields .. 207
  optical absorption spectra of ... 193
  photocatalysis product distribu-
    tion with variation in
    alkene:silane ratio,
    variation in 198
  photocatalyzed
    isomerization vs. hydrosilation 200
    reaction of $HSiEt_3$ and
      cis-2-pentene 198
    reaction of 1-pentene and
      $HSiEt_3$ 196, 197
  reactions, observed reaction
    quantum yields for 201

$M_3(CO)_{12}$ (continued)
  photochemistry 201
  spectral properties of 194
$M(CNIph)_6$ complexes,
  $M(CNPh)_6$ and 45
*ML_3^{2+} with inorganic reductants,
  rate constants for reaction of . 15
ML_6 complexes, absorption and
  emission spectral data for 48
$M(NH_3)_5L^{2+}$ complexes, comparison of spectra of 74
$Mo(N_2)_2(depe)_2$, electronic
  absorption spectrum of 151
$Mo(N_2)_2(dppe)_2$, electronic
  absorption spectrum of 150
$MoBr(N_2C_4H_9)(dppe)_2$, electronic absorption spectrum of . 153
$MoI(N_2C_8H_{17})(dppe)_2BF_4$, electronic absorption spectrum of . 155
Maxima, lifetimes, and excited-state
  energies for hydrophobic
  Ru(II) complexes,
  luminescence 33
Mechanisms, sensitization 158
Metal
  –carbonyl catalyst precursors,
    photocatalyzed reactions of
    alkenes with silanes using
    trinuclear 189
  compounds containing ligands
    with delocalized π
    systems 169–171
  compounds to sensitize an energy
    storage reaction, use of
    transition 158
  –isocyanide complexes 44
  –metal bonds 191
Methylenecyclopropanes 182
Molybdenum–carbonyl halides ... 116
Molybdenum complexes 125

N

$NaHCO_3$, photolysis of $Ru(bpy)_3^{2+}$
  in 0.01M 61
NaOH, photolysis of $Ir(bpy)_3^{2+}$
  in 0.01M 65
Nitrogens, pyridine 52
Norbornadiene (NBD)–quadri-
  cyclene(Q) interconversion 160–162
Norbornene dimers 181

O

Olefin
  complexes 174
  isomerization 139
  photoreactions by transition metal
    salts, catalysis of 174
-Osmium(II), properties of the
  excited states of tris(2,2'-
  bipyridine) ruthenium(II) and 2

SUBJECT INDEX

Os(II), properties and reactivities of the luminescent excited states of polypyridine complexes of Ru(II) and 1
Oxidation
  of Cu_{aq}^+ 21
  of Fe_{aq}^{2+} 20–22
  and reduction reactions, comparison of $*Ru(bpy)_3^{2+}$ ... 17
Oxidative quenching 12

P

1-Pentene with $HSiEt_3$, comparison of $Fe(CO)_5$- and $Fe_3(CO)_{12}$-photocatalyzed reaction of 196
1-Pentene and $HSiEt_3$, $(M_3(CO)_{12}$. photocatalyzed reaction of . 196, 197
Pentene isomerization, $Fe_3(CO)_{12}$- photocatalyzed 195
cis-2-Pentene, $M_3(CO)_{12}$-photocatalyzed reaction of $HSiEt_3$ and 198
pH, sensitized photocurrent at TiO_2 vs. 8
Phosphorescence of $Cr(bpy)_3^{3+}$, excitation spectrum (uncorrected) of the 98
Photoaquation reactions of $Ru(NH_3)_5L^{2+}$ 84
Photocatalysis 191
  product distribution with variation in the alkene:silane ratio, variation in $M_3(CO)_{12}$ 198
  quantum yields 200
Photocatalysts 189
Photocatalyzed reactions of alkenes with silanes using trinuclear metal carbonyl catalyst precursors 189
Photochemical
  aspects of the preparation of alkyldiazenido complexes .. 149
  constants for ^2E states in concentrated ClO_4^- and Cl^- aqueous solutions, photophysical and 106
  constants for ^2E states in water, photophysical and 104
  energy storage cycle 158
  processes in cyclopentadienyl-metal carbonyl complexes . 115
  reactions, sunlight-induced 158
Photochemistry 142
  of bis(dinitrogen)bis[1,2-bis(diphenylphosphino)ethane]-molybdenum 147
  of polypyridyl complexes of Cr(III), solution medium effects of the photophysics and 91
  of Ru(II) amine complexes, roles of charge transfer states in the 73

Photochromism, electron transfer . 86
Photocupration step 182
Photocycloaddition, intramolecular 179
Photodissociation 177
Photolysis
  of $CH_3CCo_3(CO)_9$ 140
  continuous 95, 102
  conventional flash 31
  of 1,5-cyclooctadiene 176
  flash 95, 100
  of $HCCo_3(CO)_9$ 137
  of $HFeCo_3(CO)_{12}$ and $HFeCo_3(CO)_{10}(PPh_3)_2$ .. 140
  of $Ir(bpy)_3^{2+}$ in 0.01M NaOH .. 65
  of $Ru(bpy)_3^{2+}$ 62, 63, 70
    and $Ir(bpy)_3^{3+}$ solutions 60
    in 1M HCl 63
    in 0.01M $NaHCO_3$ 61
  of $Ru_3(CO)_{12}$ 133
  of $Ru(NH_3)_5L^{2+}$ 86
  of $Ru(NR_3)_5L^{2+}$ 85
Photooxidation of $Rh_2(bridge)_4^{2+}$ in HX solutions, quantum yields for the 54
Photophysical and photochemical constants for ^2E states in concentrated ClO_4^- and Cl^- aqueous solutions 106
Photophysical and photochemical constants for ^2E states in water 104
Photophysics of polypyridyl complexes 91
Photoreaction of $Cr(bpy)_3^{3+}$, quantum yields for 103
Photorearrangement, rhodium-catalyzed 175
Polymer anchoring of sensitizers .. 161
Polypyridine complexes, quenching by 19
Polypyridine complexes of Ru(II) and Os(II), properties and reactivities of the luminescent excited states of 1
Polypyridineruthenium(II) complexes, energy levels in TiO_2 and 6
Polypyridyl complexes, photochemistry and photophysics of 91
Purification of materials, preparation and 30
Pyridine nitrogens 52
Pyridine solution of $Cr(CNIph)_6$, degassed 47

Q

Quadricyclene(Q) interconversion, norbornadiene(NBD)– 160
Q interconversion in the presence of triplet sensitizers, quantum yields for the NBD– 162
Q production in the presence of transition metal compounds containing ligands with delocalized π systems 170, 171

Quantum yields 66
  for CuCl-sensitized production
    of Q 165
  $M_3(CO)_{12}$ disappearance 207
  for $M_3(CO)_{12}$-photocatalyzed
    reactions, observed reaction 201
  for the NBD–Q interconversion
    in the presence of triplet
    sensitizers 162
  photocatalysis 200
  for the photooxidation of
    $Rh_2(bridge)_4^{2+}$ in HX
    solutions 54
  for photoreaction of $Cr(bpy)_3^{3+}$ 103
  for the reaction $CpFe(CO)_2X$
    + $PPh_3 \to CpFe(CO)$-
    $(PPh_3)X + CO$ 120
  for the reaction $CpMO(CO)_3X$
    + $PPh_3 \to$ cis-$CpMo$-
    $CpMo(CO)_2(PPh_3)X$ 126
  for the reactions of
    $CpMo(CO)_2(PPh_3)X$ 126
  of $Ru(bpy)_3^{2+}$, photochemical . 66
Quenching
  and the behavior of $Ru(bpy)_3^+$,
    reductive 15
  of hydrophobic ruthenium
    complexes 35–37
  oxidative 12
  by polypyridine complexes 19
  and reaction with neutral
    electron donors 38
  of Ru(II) complex excited states
    by amines 38

R

$Rh_2(bridge)_4^{2+}$ 52
  in HX solutions, quantum yields
    for the photoooxidation of .. 54
Ru(II)
  amine complexes, roles of charge
    transfer states in the
    photochemistry of 73
  2,2'-bipyridine complexes of
    Ir(III) and 57
  complex excited states by
    amines, quenching of 38
  complexes, luminescence maxima,
    lifetimes, and excited state
    energies for hydrophobic ..33–37
  and Os(II), properties and reac-
    tivities of the luminescent
    excited states of polypyri-
    dine complexes of 1
$Ru(bpy)_3^+$, reductive quenching
    and the behavior of 15
$Ru(bpy)_3^{2+}$
  absorption spectrum of 2
  double minima potential wells for 70
  energy transfer in 66
  free energy diagram for the
    deactivation of the lumines-
    cent excited state of 9

$Ru(bpy)_3^{2+}$ (continued)
  in 1M HCl, photolysis of 63
  and $Ir(bpy)_3^{3+}$ solutions,
    photolysis of 60
  luminescent excited state of ... 5
  in 0.01M $NaHCO_3$, photolysis of 61
  photochemical quantum yields of 66
  photolysis of 62, 63, 70
*$Ru(bpy)_3^{2+}$ oxidation and reduc-
    tion reactions, comparison of . 17
*$Ru(bpy)_3^{2+}$ and $Ru(bpy)_3^+$, rate
    constants for reduction by ... 15
$Ru(bipy)_3^{2+}$, light-induced elec-
    tron transfer reactions of
    hydrophobic analogs of 28
$Ru_2(CO)_{12}$, photolysis of 133
$Ru(NH_3)_6^{2+}$ 77–80
  electron spectrum of 75
$Ru(NH_3)_5L^{2+}$, photoaquation
    reactions of 84
$Ru(NH_3)_5L^{2+}$, photolysis of 86
$Ru(NH_3)_5py^{2+}$, electronic
    spectrum of 76
$Ru(NH_3)_5(py-x)^{2+}$ 81
  complexes 83
$Ru(NR_3)_5L^{2+}$ 85
Rate constants
  for outer–sphere reductions by
    Cr_{aq}^{2+} and Eu_{aq}^{2+} 13
  for reaction of *ML_3^{2+} with
    inorganic reductants 15
  for reduction by *$Ru(bpy)_3^{2+}$
    and $Ru(bpy)_3^+$ 15
Reaction of *ML_3^{2+} with inorganic
    reductants, rate constants for 15
Reactivity, excited state 11
Rearrangement, allylic 175
Redox couples for hydrophobic
    ruthenium complexes 34
Reduction(s)
  by Cr_{aq}^{+2} and Eu_{aq}^{2+}, rate
    constants for outer-sphere . 13
  potentials, excited-state 7
  quenching and the behavior of
    $Ru(bpy)_3^+$ 15
  reactions, comparison of
    *$Ru(bpy)_3^{2+}$ oxidation and 17
  by *$Ru(bpy)_3^{2+}$ and $Ru(bpy)_3^+$,
    rate constants for 15
Rhodium 174
  -catalyzed photorearrangement . 175
  dihydride 178
Ring expansion 183

S

Salts, catalysts of olefin photoreac-
    tions by transition metal 174
Sensitization
  by copper(I) compounds 163
  mechanisms 158
  by transition metal compounds
    containing ligands with
    delocalized π systems 169

SUBJECT INDEX

Sensitizer(s)
  development 160
  polymer anchoring of 161
  quatum yields for the NBD–Q
    interconversion in the
    presence of triplet 162
Seven-coordinate intermediate ... 91
Silane(s)
  ratio, product distribution variation with alkene– 197
  ratio, variation in $M_3(CO)_{12}$
    photocatalysis product distribution with variation in
    alkene– 198
  reaction of alkenes with 195
  using trinuclear metal carbonyl
    catalyst precursors, photocatalyzed reactions of
    alkenes with 189
Solar energy 158
  conversion 44
Solution medium 94
  effects on the photophysics and
    photochemistry of polypyridyl complexes of Cr(III) .. 91
Solvents, effect of 107
Spectral
  data for $CpMo(CO)_n$-
    $(PPh_3)_{3-n}X$ complexes 125
  data for ML_6 complexes, absorption and emission 48
  measurements 136
  properties of $M_3(CO)_{12}$ 194
  sensitizer 160
Spectroscopy 31
  ground state absorption 95
  luminescence 96
Spectrum(a)
  absorption 94
  of $Cr(bpy)_3^{3+}$ 96, 97
  electronic 136–141
    of $CH_3CCo_3(CO)_9$ 136
    of $CpFe(CO)_2Br$ 119
    of $CpFe(CO)_2Cl$ 119
    of $CpFe(CO)_2I$ 119
    of $Cu[P(C_6H_5)_3]_2BH_4$ and
      $P(C_6H_5)_3$ 168
    of dinitrogen, alkyldiazenido, and alkylhydrazido derivatives of
      molybdenum 150
    of $MoBr(N_2C_4H_9)(dppe)_2$ 153
    of $MoI(N_2C_8H_{17})$-
      $(dppe)_2BF_4$ 155
    of $Mo(N_2)_2(depe)_2$ 151
    of $Mo(N_2)_2(dppe)_2$ 150
    of $Ru(NH_3)_6^{2+}$ 75
    of $Ru(NH_3)_5py^{2+}$ 76
    of $M_3(CO)_{12}$, optical 193

Spectrum(a) (continued)
  emission, of $Cr(bpy)_3^{3+}$,
    phosphorescence 98
  emission, of $Ru(bpy)_3^{2+}$ 69
  excited state 3
  types of 74
  IR, of $CpFe(CO)_2NCS$ 124
  IR, of $CpFe(CO)_2SCN$ 124
  of $M(NH_3)_5L^{2+}$ complexes,
    comparison of 74
  of the phosphorescence of
    $Cr(bpy)_3^{3+}$, excitation 98
Sunlight-induced photochemical
  reactions 158

T

Thermal reactivity of CH_3CCo_3-
  $(CO)_8PPh_3$ and CH_3CCo_3-
  $(CO)_7(PPh_3)_2$ 141
TiO_2 and polypyridineruthenium(II) complexes, energy
  levels in 6
TiO_2 vs. pH, sensitized photocurrent at 8
Time, plots of absorbance against
  irradiation 204, 205
Time, plots of absorbance against
  355-nm irardiation 206
Transfer
  hydrogen 178
  hydrogenation 175
  states in the photochemistry of
    Ru(II) amine complexes,
    roles of charge 73
Transition metal
  compounds containing ligands
    with delocalized π
    systems 169–171
  compounds to sensitize an
    energy storage reaction,
    use of 158
  salts catalysis of olefin photoreactions by 174
Trinuclear clusters 192
Triphenylphosphine 122

W

$W(CNPh)_6$, degassed $CHCl_3$
  solution of 50
Water, photophysical and photochemical constants for ^2E
  states in 104

Y

Yield determinations, quantum ... 31

The text of this book is set in 10 point Caledonia with two points of leading. The chapter numerals are set in 30 point Garamond; the chapter titles are set in 18 point Garamond Bold.

The book is printed offset on Text White Opaque 50-pound. The cover is Joanna Book Binding blue linen.

Jacket design by Alan Kahan. Editing and production by Saundra Goss.

The text was composed by Service Composition Co., Baltimore, MD; the structures were set by the Mack Printing Co., Easton, PA; and the book was printed and bound by The Maple Press Co., York, PA.